LAND SUBSIDENCE IN THE UNITED STATES

EDITED BY

Devin Galloway

David R. Jones

S.E. Ingebritsen

U.S. Geological Survey

Circular 1182

1999

U.S. DEPARTMENT OF THE INTERIOR

Bruce Babbitt
Secretary

U.S. GEOLOGICAL SURVEY

Charles G. Groat
Director

Reston, Virginia 1999

Free on application to the

U.S. Geological Survey
Information Services
Box 25286
Denver, CO 80225-0286

Library of Congress Cataloging-in-Publication Data

Land subsidence in the United States / edited by Devin Galloway, David R. Jones, S.E. Ingebritsen.
    p. cm. — (U.S. Geological Survey Circular; 1182)
    Includes bibliographical references.
    ISBN 0-607-92696-1
    1. Subsidence (Earth movements)—United States—Case studies.
    2. Groundwater—United States—Case studies I. Galloway, Devin L.
    II. Jones, David R. (David Richard) III. Ingebritsen, S.E. IV. Series.

QE600.3.U6.L36 1999
551.3´07´0973—dc21   99-040089

Foreword

Sacramento/San Joaquin River Delta

From the San Francisco Bay/Delta to the Florida Everglades and from upstate New York to Houston, people are dealing with a common problem in these diverse locations—land subsidence due to the withdrawal of ground water or the application of water at the land surface. These locations illustrate that subsidence is not an isolated problem: an area of more than 15,000 square miles in 45 States experience land subsidence. Using these locations and others as case studies, this report focuses on three principal processes causing land subsidence: the compaction of aquifer systems, the oxidation of organic soils, and the collapse of cavities in carbonate and evaporite rocks. The impacts of land subsidence, past and present, are illustrated, and most importantly, so is the value of science in effectively limiting damages from land subsidence.

An important aspect of the USGS mission is to provide information that describes the Earth, its resources, and the processes that govern the availability and quality of those resources. With reports such as this Circular, the USGS seeks to broaden public understanding of land subsidence as an Earth process, and the serious impacts that subsidence can cause if those impacts are not understood, anticipated, and properly managed. By applying scientific understanding and engineering approaches to problems of land subsidence, our society will have solutions that can mitigate or eliminate the negative impacts of subsidence while allowing continued beneficial uses of water. It is our hope that this information will be helpful for concerned citizens, landowners, water users, water managers, and officials responsible for public investments and regulation of land and water use.

For some readers, this report will be an end in itself in providing an understanding of the phenomena of land subsidence that satisfies their need to act as informed citizens or decision makers, or simply to satisfy their curiosity about an important Earth process. For other readers, we hope this report will be a gateway to the rich scientific literature on the subject of subsidence and strategies for the control of subsidence, through the references provided.

Scientific understanding is critical to the formulation of balanced decisions about the management of land and water resources. This Circular coupled with ongoing data collection, basic research, and applications of that research to specific subsidence problems, constitute the USGS contribution toward wise management of land subsidence as a part of effective and publicly beneficial land- and water-management strategies.

Robert M. Hirsch
Associate Director for Water Resources

Fissure, South-Central Arizona

Acknowledgments

We are particularly indebted to four U.S. Geological Survey colleagues who assisted in the planning and research that led to this report. Keith Prince, Stan Leake and Tom Holzer, long-time proponents of the scientific understanding of subsidence problems related to ground-water use, launched this effort. Francis Riley, a pioneer in field studies of land subsidence related to ground-water extraction, counselled us extensively on aquifer-system compaction.

We are also especially grateful to the many key colleagues and cooperators who generously lent their expertise to review technical and nontechnical aspects of each of the case studies: Behzad Ahmadi, Tom Iwamura, and Cheryl Wessling (Santa Clara Valley Water District), and Eric Reichard (U.S. Geological Survey) for Santa Clara Valley, California; Gil Bertoldi (U.S. Geological Survey, retired), George Davis (U.S. Geological Survey, retired) and Harvey Swanson (California Division of Water Resources, retired) for San Joaquin Valley, California; Robert Gabrysch (U.S. Geological Survey, retired), and Ron Neighbors (Harris-Galveston Coastal Subsidence District) for Houston-Galveston, Texas; John Bell (Nevada Bureau of Mines and Geology), Gary Dixon (U.S. Geological Survey), and Michael Johnson (Las Vegas Valley Water District) for Las Vegas Valley, Nevada; Stan Leake (U.S. Geological Survey), and Herb Schumann (U.S. Geological Survey, retired) for Southern Arizona; Margit Aramburu (Delta Protection Commission), Lauren Hastings (U.S. Geological Survey), and M. Mirmazaheri (California Department of Water Resources) for the Sacramento-San Joaquin River Delta, California; Jud Harvey (U.S. Geological Survey), Carol Kendall (U.S. Geological Survey), and Jayantha Obeysekera (South Florida Water Management District) for the Florida Everglades; Jim Borchers (U.S. Geological Survey), Kathy Sanford (New York State Department of Environmental Conservation) and Richard Young (State University New York—Geneseo) for the Retsof Salt Mine Collapse, New York; and Mark Barcelo (Southwest Florida Water Management District), Craig Hutchinson (U.S. Geological Survey), William L. Wilson (Subsurface Evaluations Inc.), and Dan Yobbi (U.S. Geological Survey) for West-Central Florida. We are also grateful to our U.S. Geological Survey colleagues, Charles Heywood and Steven Phillips, for thorough and thoughtful reviews of the final chapter, Role of Science. Finally we thank Michelle Sneed (U.S. Geological Survey) for her final read-through of the Circular and for her constructive comments.

Cover-collapse sinkhole, West-Central Florida

Contents

Conversions

This Circular uses English units. To determine metric values use the conversion factors listed below.

MEASUREMENT	MULTIPLY	BY	TO OBTAIN
Length	inch	25.4	millimeter
	foot	0.3048	meter
	mile	1.609	kilometer
Area	square foot	0.09290	square meter
	square mile	2.590	square kilometer
	acre	0.4047	hectare
Volume	acre foot	1233	cubic meter
	cubic foot	0.02832	cubic meter
	gallon	3.785	liter
Mass	ounce	28.35	gram
	pound	0.4536	kilogram
	ton (short)	0.9072	megagram
Temperature	degree Fahrenheit	$\dfrac{°F-32}{1.8}$	degree Celsius

Vertical Datum

In this report, "sea level" refers to the National Geodetic Vertical Datum of 1929 (NGVD of 1929)—a geodetic datum derived from a general adjustment of the first-order level nets of both the United States and Canada, formerly called "Sea Level Datum of 1929." "Mean sea level" is not used with reference to any particular vertical datum; where used, the phrase means the average surface of the ocean as determined by calibration of measurements at tidal stations.

INTRODUCTION

Land subsidence in the United States

This earth fissure formed as a result of differential compaction of the aquifer system near Mesa, Arizona.

L and subsidence is a gradual settling or sudden sinking of the Earth's surface owing to subsurface movement of earth materials. Subsidence is a global problem and, in the United States, more than 17,000 square miles in 45 States, an area roughly the size of New Hampshire and Vermont combined, have been directly affected by subsidence. The prinicipal causes are aquifer-system compaction, drainage of organic soils, underground mining, hydrocompaction, natural compaction, sinkholes, and thawing permafrost (National Research Council, 1991). More than 80 percent of the identified subsidence in the Nation is a consequence of our exploitation of underground water, and the increasing development of land and water resources threatens to exacerbate existing land-subsidence problems and initiate new ones. In many areas of the arid Southwest, and in more humid areas underlain by soluble rocks such as limestone, gypsum, or salt, land subsidence is an often-overlooked environmental consequence of our land- and water-use practices.

In 1991, the National Research Council estimated that annual costs in the United States from flooding and structural damage caused by land subsidence exceeded $125 million. The assessment of other costs related to land subsidence, especially those due to ground-water withdrawal, is complicated by difficulties in identifying and mapping the affected areas, establishing cause-and-effect relations, assigning economic value to environmental resources, and by inherent conflicts in the legal system regarding the recovery of damages caused by resource removal under established land and water rights. Due to these "hidden" costs, the total cost of subsidence is probably significantly larger than our current best estimate.

We explore the role of underground water in human-induced land subsidence through illustrative case histories. Extraction and drainage of ground water play direct roles in land subsidence by causing the compaction of susceptible aquifer systems and the dewatering of organic soils. The catastrophic formation of sinkholes in susceptible earth materials, although fundamentally a natural process, can

Subsidence occurs worldwide
Three famous examples of subsidence

AQUIFER-SYSTEM COMPACTION IN MEXICO CITY

In Mexico City, rapid land subsidence caused by ground-water withdrawal and associated aquifer-system compaction has damaged colonial-era buildings, buckled highways, and disrupted water supply and waste-water drainage. Maximum rates of subsidence approach 2 feet per year and total subsidence during the 20th century is as great as 30 feet (New York Times International, January 29, 1998). In the downtown area, the steel casings of wells drilled deep enough to penetrate beneath the subsiding aquifer system now protrude 20 feet or more above ground. The progressive sinking of the urban area has rendered the original waste-water drainage system ineffective, and forced construction of a new, deep, 124-mile-long sewer network.

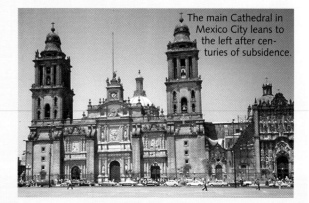

The main Cathedral in Mexico City leans to the left after centuries of subsidence.

ORGANIC-SOIL SUBSIDENCE AND THE DUTCH LANDSCAPE

It is said that "God created the world, but the Dutch created Holland." Near-sea-level marshlands in the western Netherlands began to be drained for agriculture between the 9th and 14th centuries, and by the 16th century the land had subsided to the extent that windmills were needed to artificially discharge water to the sea. The classic Dutch landscape of dikes, canals, and windmills reflects centuries of reclamation and consequent subsidence. Average subsidence rates have increased during the 20th century because of greatly improved drainage.

DISSOLUTION-COLLAPSE FEATURES ON THE YUCATAN PENINSULA

The low-lying Yucatan Peninsula of eastern Mexico is covered by a blanket of limestone, and dissolution of the limestone by infiltrating rainwater has created a highly permeable aquifer, comparable to the Floridan aquifer of the Florida peninsula. Infiltration of rainwater is so rapid that there are no surface streams. For millennia, human civilizations relied on sinkholes formed by collapse of rock above subsurface cavities—locally known as cenotes—for water supply. Great troves of Mayan relics have been found in some cenotes.

Cenote at Chichén Itzá, Mexico

(Clive Ruggles, Leicester University, UK, 1986)

During the construction of a railroad northeast of Valdez, Alaska, the permafrost's thermal equilibrium was disrupted, causing differential thawing that warped the roadbed. The railroad was abandoned in 1938, but subsidence has continued.

also be triggered by ground-water-level declines caused by pumping, or by infiltration from reservoir impoundments, surface-water diversions, or storm runoff channels. The case histories illustrate the three basic mechanisms by which human influence on ground water causes land subsidence—compaction of aquifer systems, dewatering of organic soils, and mass wasting through dissolution and collapse of susceptible earth materials. We also examine the role that science and water-management groups play in mitigating subsidence damages.

Several other types of subsidence involve processes more or less similar to the three mechanisms just cited, but are not covered in detail in this Circular. These include the consolidation of sedimentary deposits on geologic time scales; subsidence associated with tectonism; the compaction of sediments due to the removal of oil and gas reserves; subsidence of thawing permafrost; and the collapse of underground mines. Underground mining for coal accounts for most of the mining-related subsidence in the United States and has been thoroughly addressed through Federal and State programs prompted by the 1977 Surface Mining Control and Reclamation Act. No such nationally integrated approach has been implemented to deal with the remaining 80 percent of land subsidence associated with ground-water processes.

Oil and gas removal in Long Beach, California caused subsidence. Levees were built to prevent flooding of the oil fields and port facilities.

Subsidence pits and troughs formed above the Dietz coal mines near Sheridan, Wyoming. The coal mines were in operation from the 1890s to the 1920s.

An undeveloped aquifer system is in balance between recharge and discharge. Pumping for urban or agricultural uses disrupts this balance and may cause subsidence to occur.

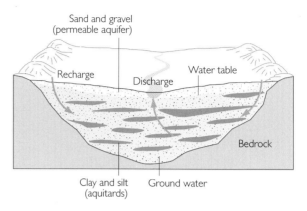

Mining ground water We begin with five case histories in which overdraft of susceptible aquifer systems has resulted in regional, permanent subsidence and related ground failures. In alluvial aquifer systems, especially those that include semiconsolidated silt and clay layers (aquitards) of sufficient aggregate thickness, long-term ground-water-level declines can result in a vast one-time release of "water of compaction" from compacting aquitards, which manifests itself as land subsidence. Accompanying this release of water is a largely nonrecoverable reduction in the pore volume of the compacted aquitards, and thus an overall reduction in the total storage capacity of the aquifer system. This "water of compaction" cannot be reinstated by allowing water levels to recover to their predevelopment status. The extraction of this resource for economic gain constitutes ground-water mining in the truest sense of the term.

The five case studies demonstrate how agricultural and municipal-industrial ground-water use have combined to deplete critical ground-water resources and create costly regional-scale subsidence. We begin in the "Silicon Valley" in northern California, where early agricultural ground-water use contributed to subsidence that has increased flood risks in the greater San Jose area. Silicon Valley (properly the Santa Clara Valley) was the first place in the United States where subsidence due to ground-water pumpage was recognized; since the late 1960s, the ground-water resource there has been successfully managed to halt subsidence. In nearby San Joaquin Valley, the single largest human alteration of the Earth's surface topography resulted from excessive ground-water pumpage to sustain an exceptionally productive agriculture. In the Houston-Galveston area in Texas, early production of oil and gas, and a long history of ground-water pumpage, have created severe and costly coastal-flooding hazards and affected a critical environmental resource—the Galveston Bay estuary. In Las Vegas Valley ground-water depletion and subsidence have accompanied the conversion of a desert oasis into a thirsty and fast-growing metropolis. Finally, in south-central Arizona, importation of Colorado River water and conversion of water-intensive agriculture to lower-water-demand urban land uses has helped to partly arrest subsidence and forestall further fissuring of the Earth's surface.

The organic soils of the Florida Everglades are quickly disappearing.

Drainage of organic soils Land subsidence invariably occurs when organic soils—soils rich in organic carbon—are drained for agriculture or other purposes. The most important cause of this subsidence is microbial decomposition which, under drained conditions, readily converts organic carbon to carbon-dioxide gas and water. Compaction, desiccation, erosion by wind and water, and prescribed or accidental burning can also be significant factors.

The total area of organic soils in the United States is roughly equivalent to the size of Minnesota, about 80,000 square miles, nearly half of which is "moss peat" located in Alaska (Lucas, 1982). About 70 percent of the organic-soil area in the contiguous 48 states occurs in northerly, formerly glaciated areas, where moss peats are also common (Stephens and others, 1984). Moss peat is composed mainly of sphagnum moss and associated species. It is generally very acidic (pH 3.5 to 4) and, therefore, not readily decomposed, even when drained. However, where moss peat is amended for agricultural cultivation, for example through fertilization and heavy application of lime to raise the pH, it can decompose nearly as rapidly as other types of organic soils.

Our two case studies of organic-soil subsidence focus on examples of rapid subsidence (1 to 3 inches/year) caused by decomposition of the remains of shallow-water sedges and reeds. In the Sacramento-San Joaquin Delta of California and the Florida Everglades, continuing organic-soil subsidence threatens agricultural production, affects engineering infrastructure that transfers water supplies to large urban populations, and complicates ongoing ecosystem-restoration efforts sponsored by the Federal and State governments.

Collapsing cavities The final two case studies deal with the sudden and sometimes catastrophic land subsidence associated with localized collapse of subsurface cavities—sinkholes. This type of subsidence is commonly triggered by ground-water-level declines caused by pumping and by enhanced percolation of water through susceptible rocks. Collapse features tend to be associated with specific rock types having hydrogeologic properties that render them susceptible to dissolution in water and the formation of cavities. Evaporite minerals (salt, gypsum and anhydrite) and carbonate minerals (limestone and dolomite) are susceptible to extensive dissolution by water. Salt and gypsum are, respectively, almost 7,500 and 150 times more soluble than limestone, the rock type often associated with catastrophic sinkhole formation.

Evaporite rocks underlie about 35 to 40 percent of the United States, although in many areas at depths so great as to have no discernible effect at land surface. Natural solution-related subsidence has occurred in each of the major salt basins (Ege, 1984) throughout the United States. The high solubilities of salt and gypsum permit cavities to form in days to years, whereas cavity formation in carbonate bedrock is a very slow process that generally occurs over centuries to millennia. The slow dissolution of carbonate rocks favors the stability and persistence of the distinctively weathered landforms known as karst. Carbonate karst landscapes comprise more than 40 percent

Cover collapse sinkhole in Winter Park, Florida, 1981

of the humid United States east of the longitude of Tulsa, Oklahoma (White and others, 1995). Human activities can facilitate the formation of subsurface cavities in these susceptible materials and trigger their collapse, as well as the collapse of pre-existing subsurface cavities. Though the collapse features tend to be highly localized, their impacts can extend beyond the collapse zone via the potential introduction of contaminants to the ground-water system. Our two cavity-collapse case studies—Retsof, New York and west-central Florida—focus on human-induced cavity collapses in salt and limestone, respectively.

The role of science In a final section we discuss the role of science in defining subsidence problems and understanding subsidence processes. A combination of scientific understanding and careful management can minimize the subsidence that results from developing our land and water resources.

PART I

Mining Ground Water

Santa Clara Valley, California

San Joaquin Valley, California

Houston-Galveston, Texas

Las Vegas, Nevada

South-Central Arizona

Permanent subsidence can occur when water stored beneath the Earth's surface is removed by pumpage or drainage. The reduction of fluid pressure in the pores and cracks of aquifer systems, especially in unconsolidated rocks, is inevitably accompanied by some deformation of the aquifer system. Because the granular structure—the so-called "skeleton"—of the aquifer system is not rigid, but more or less compliant, a shift in the balance of support for the overlying material causes the skeleton to deform slightly. Both the aquifers and aquitards that constitute the aquifer system undergo deformation, but to different degrees. Almost all the permanent subsidence occurs due to the irreversible compression or consolidation of aquitards during the typically slow process of aquitard drainage (Tolman and Poland, 1940). This concept, known as the aquitard-drainage model, has formed the theoretical basis of many successful subsidence investigations.*

Areas where subsidence has been attributed to ground-water pumpage

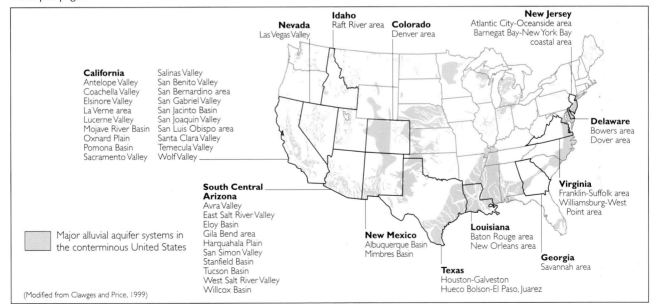

California
Antelope Valley
Coachella Valley
Elsinore Valley
La Verne area
Lucerne Valley
Mojave River Basin
Oxnard Plain
Pomona Basin
Sacramento Valley

Salinas Valley
San Benito Valley
San Bernardino area
San Gabriel Valley
San Jacinto Basin
San Joaquin Valley
San Luis Obispo area
Santa Clara Valley
Temecula Valley
Wolf Valley

Nevada
Las Vegas Valley

Idaho
Raft River area

Colorado
Denver area

New Jersey
Atlantic City-Oceanside area
Barnegat Bay-New York Bay
coastal area

Delaware
Bowers area
Dover area

Virginia
Franklin-Suffolk area
Williamsburg-West
Point area

South Central Arizona
Avra Valley
East Salt River Valley
Eloy Basin
Gila Bend area
Harquahala Plain
San Simon Valley
Stanfield Basin
Tucson Basin
West Salt River Valley
Willcox Basin

New Mexico
Albuquerque Basin
Mimbres Basin

Louisiana
Baton Rouge area
New Orleans area

Georgia
Savannah area

Texas
Houston-Galveston
Hueco Bolson-El Paso, Juarez

Major alluvial aquifer systems in the conterminous United States

(Modified from Clawges and Price, 1999)

---

* Studies of subsidence in the Santa Clara Valley (Tolman and Poland, 1940; Poland and Green, 1962; Green, 1964; Poland and Ireland, 1988) and San Joaquin Valley (Poland, 1960; Miller, 1961; Riley, 1969; Helm, 1975; Poland and others, 1975; Ireland and others, 1984) in California established the theoretical and field application of the laboratory derived principle of effective stress and theory of hydrodynamic consolidation to the drainage and compaction of aquitards. For reviews of the history and application of the aquitard drainage model see Holzer (1998) and Riley (1998).

When water levels drop, due mainly to seasonal increases in ground-water pumping, some support for the overlying material shifts from the pressurized fluid filling the pores to the granular skeleton of the aquifer system.

When ground water is recharged and water levels rise, some support for the overlying material shifts from the granular skeleton to the pressurized pore fluid.

Land surface

Unconfined aquifer

Sand and gravel

The increased load compresses the skeleton by contracting the pore spaces, causing some lowering of the land surface.

Confined aquifer

Clay and silt (aquitards)

Depth to water

Time

Land surface

Under the decreased load the pore spaces and the skeleton expand, causing some raising of the land surface.

Not to scale

Contracting aquifer-system skeleton

Expanding aquifer-system skeleton

Pore space

Clay particle

Decreased fluid pressure causes the skeleton to contract, creating some small subsidence of land surface.

Increased fluid pressure expands the skeleton, creating some small uplift of land surface.

REVERSIBLE DEFORMATION OCCURS IN ALL AQUIFER SYSTEMS

The relation between changes in ground-water levels and compression of the aquifer system is based on the principle of effective stress first proposed by Karl Terzaghi (Terzaghi, 1925). By this principle, when the support provided by fluid pressure is reduced, such as when ground-water levels are lowered, support previously provided by the pore-fluid pressure is transferred to the skeleton of the aquifer system, which compresses to a degree. Conversely, when the pore-fluid pressure is increased, such as when ground water recharges the aqui-

Mostly recoverable (elastic) deformation was observed during and following a pumping test near Albuquerque, New Mexico. Changes in the water level due to cyclic pumping were accompanied by alternating cycles of compression and expansion of the aquifer system.

A measure of the change in applied stress is the change in water level.

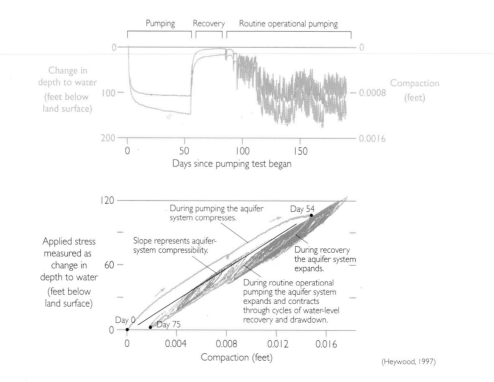

Change in depth to water (feet below land surface)

Pumping   Recovery   Routine operational pumping

Compaction (feet)

Days since pumping test began

Applied stress measured as change in depth to water (feet below land surface)

During pumping the aquifer system compresses.

Slope represents aquifer-system compressibility.

Day 54

During recovery the aquifer system expands.

During routine operational pumping the aquifer system expands and contracts through cycles of water-level recovery and drawdown.

Day 0   Day 75

Compaction (feet)

(Heywood, 1997)

fer system, support previously provided by the skeleton is transferred to the fluid and the skeleton expands. In this way, the skeleton alternately undergoes compression and expansion as the pore-fluid pressure fluctuates with aquifer-system discharge and recharge. When the load on the skeleton remains less than any previous maximum load, the fluctuations create only a small elastic deformation of the aquifer system and small displacement of land surface. This fully recoverable deformation occurs in all aquifer systems, commonly resulting in seasonal, reversible displacements in land surface of up to 1 inch or more in response to the seasonal changes in ground-water pumpage.

INELASTIC COMPACTION IRREVERSIBLY ALTERS THE AQUIFER SYSTEM

The maximum level of past stressing of a skeletal element is termed the preconsolidation stress. When the load on the aquitard skeleton exceeds the preconsolidation stress, the aquitard skeleton may undergo significant, permanent rearrangement, resulting in irreversible compaction. Because the skeleton defines the pore structure of the aquitard, this results in a permanent reduction of pore volume as the pore fluid is "squeezed" out of the aquitards into the aquifers. In confined aquifer systems subject to large-scale overdraft, the volume of water derived from irreversible aquitard compaction is essentially equal to the volume of subsidence and can typically range from 10 to 30 percent of the total volume of water pumped. This represents a one-time mining of stored ground water and a small permanent reduction in the storage capacity of the aquifer system.

When long-term pumping lowers ground-water levels and raises stresses on the aquitards beyond the preconsolidation-stress thresholds, the aquitards compact and the land surface subsides permanently.

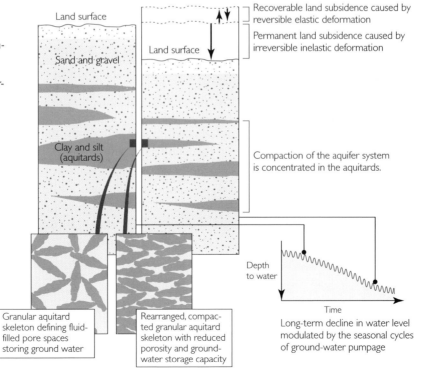

Aquitard Drainage and Aquifer-System Compaction
The Principle of Effective Stress

This principle describes the relation between changes in water levels and deformation of the aquifer system.

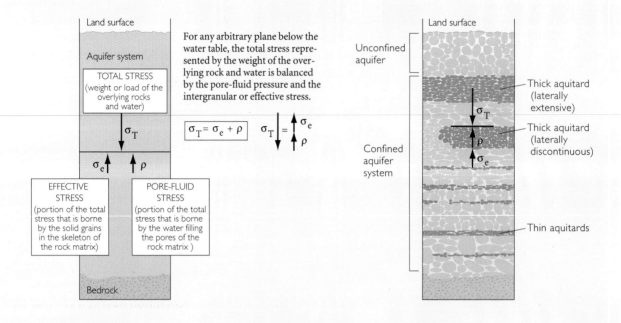

For any arbitrary plane below the water table, the total stress represented by the weight of the overlying rock and water is balanced by the pore-fluid pressure and the intergranular or effective stress.

$$\sigma_T = \sigma_e + \rho$$

$$\sigma_T = \begin{array}{c} \sigma_e \\ \rho \end{array}$$

TOTAL STRESS
(weight or load of the overlying rocks and water)

EFFECTIVE STRESS
(portion of the total stress that is borne by the solid grains in the skeleton of the rock matrix)

PORE-FLUID STRESS
(portion of the total stress that is borne by the water filling the pores of the rock matrix)

PROLONGED CHANGES IN GROUND-WATER LEVELS INDUCE SUBSIDENCE

Prior to the extensive development of ground-water resources, water levels are relatively stable—though subject to seasonal and longer-term climatic variability.

During development of ground-water resources, water levels decline and land subsidence begins.

After ground-water pumping slows or decreases, water levels stabilize but land subsidence may continue.

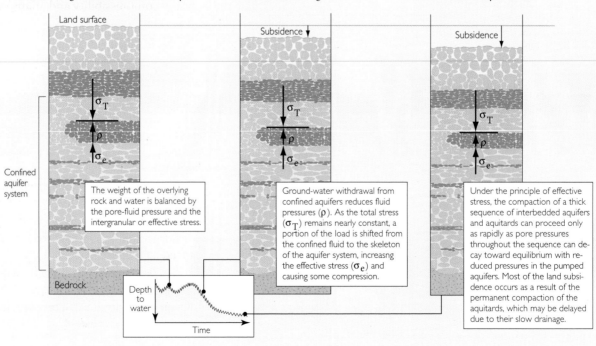

The weight of the overlying rock and water is balanced by the pore-fluid pressure and the intergranular or effective stress.

Ground-water withdrawal from confined aquifers reduces fluid pressures (ρ). As the total stress (σ_T) remains nearly constant, a portion of the load is shifted from the confined fluid to the skeleton of the aquifer system, increasng the effective stress (σ_e) and causing some compression.

Under the principle of effective stress, the compaction of a thick sequence of interbedded aquifers and aquitards can proceed only as rapidly as pore pressures throughout the sequence can decay toward equilibrium with reduced pressures in the pumped aquifers. Most of the land subsidence occurs as a result of the permanent compaction of the aquitards, which may be delayed due to their slow drainage.

More than 2.5 feet of permanent (inelastic) compaction was observed near Pixley, San Joaquin Valley, California during a 10-year period.

The high summer demand for irrigation water combined with the normally wetter winters causes ground-water levels to fluctuate in response to seasonal pumpage and recharge. The annual cycles of alternating stress increase and decrease are accompanied by cycles of compression and slight expansion of the aquifer system.

Compression proceeds most rapidly when the stress is larger than the preconsolidation stress threshold. Beyond this threshold almost all of the compression is permanent (inelastic) and attributed to the compaction of fine-grained aquitards.

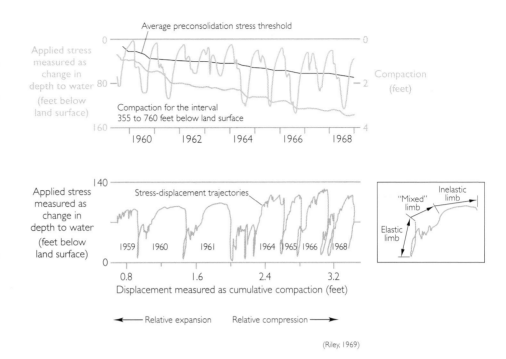

(Riley, 1969)

*"… the term **aquitard** has been coined to describe the less-permeable beds in a stratigraphic sequence. These beds may be permeable enough to transmit water in quantities that are significant in the study of regional ground-water flow, but their permeability is not sufficient to allow the completion of production wells within them."*

—Freeze and Cherry, 1979

Aquitards play an important role in compaction

In recent decades increasing recognition has been given to the critical role of aquitards in the intermediate and long-term response of alluvial aquifer systems to ground-water pumpage. In many such systems interbedded layers of silts and clays, once dismissed as non-water yielding, comprise the bulk of the ground-water storage capacity of the confined aquifer system! This is by virtue of their substantially greater porosity and compressibility and, in many cases, their greater aggregate thickness compared to the more transmissive, coarser-grained sand and gravel layers.

Because aquitards are by definition much less permeable than aquifers, the vertical drainage of aquitards into adjacent pumped aquifers may proceed very slowly, and thus lag far behind the changing water levels in adjacent aquifers. The duration of a typical irrigation season may allow only a modest fraction of the potential yield from aquitard storage to enter the aquifer system, before pumping ceases for the season and ground-water levels recover in the aquifers. Typically, for thick aquitards, the next cycle of pumping begins before the fluid pressures in the aquitards have equilibrated with the previous cycle. The lagged response within the inner portions of a thick aquitard may be largely isolated from the higher frequency seasonal fluctuations and more influenced by lower frequency, longer-term trends in ground-water levels. Because the migration of increased internal stress into the aquitard accompanies its drainage, as more fluid is squeezed from the interior of the aquitard, larger and larger internal stresses propagate farther into the aquitard.

When the internal stresses exceed the preconsolidation stress, the compressibility increases dramatically, typically by a factor of 20 to

Aquitard Drainage and Aquifer-System Compaction
The Theory of Hydrodynamic Consolidation

The theory describes the delay in draining aquitards when water levels are lowered in adjacent aquifers, as well as the residual compaction that may continue long after water levels are initially lowered.

During a 90-year period (1908–1997) of ground-water development in the Antelope Valley, California, the response of water levels in two thick aquitards lags the declining water level in the aquifer. A laterally discontinuous aquitard draining from both upper and lower faces approaches fluid-pressure equilibrium with the adjacent aquifers more rapidly than an overlying laterally extensive aquitard that has a complex drainage history, including a gradient reversal.*

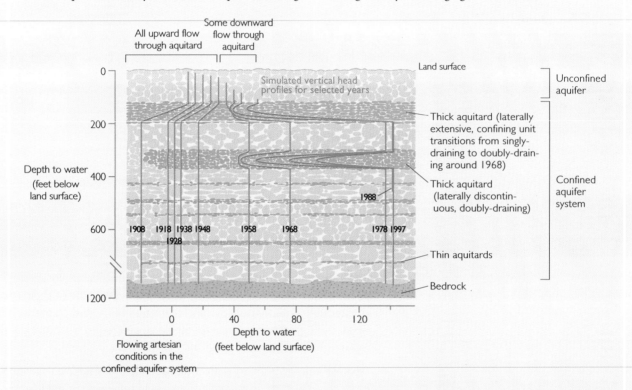

RESIDUAL COMPACTION

Significant amounts of compaction began occurring in the late 1950s after water levels in the aquifers had fallen some 60 feet. Initially, most of the compaction occured in the faster-draining thin aquitards within the aquifers. Subsequently most of the compaction occured in the two thickest and most slowly draining aquitards. Despite stabilization of ground-water levels in the aquifers, more than 0.3 feet of compaction has occurred since 1990, due to residual compaction.

Simulations predict that another 1.3 feet of compaction may ultimately occur even if ground-water levels remain at 1997 levels.

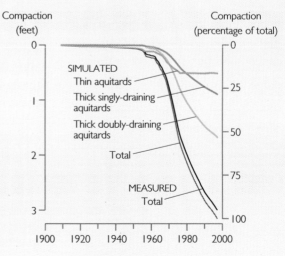

*These results from an aquifer system in Antelope Valley, Mojave Desert, California are based on field measurements and computer simulations of aquitard drainage. They illustrate the history of ground-water-level changes and compaction in the aquifers and aquitards throughout the period of ground-water resource development, 1908-97.

(Michelle Sneed, USGS. written communication, 1998)

100 times, and the resulting compaction is largely nonrecoverable. At stresses greater than the preconsolidation stress, the lag in aquitard drainage increases by comparable factors, and concomitant compaction may require decades or centuries to approach completion. The theory of hydrodynamic consolidation (Terzaghi, 1925)—an essential element of the "aquitard drainage model"—describes the delay involved in draining aquitards when heads are lowered in adjacent aquifers, as well as the residual compaction that may continue long after drawdowns in the aquifers have essentially stabilized. Numerical modeling based on Terzaghi's theory has successfully simulated complex histories of compaction observed in response to measured water-level fluctuations (Helm, 1978).

Hydrodynamic lag, which is a delay in the propagation of fluid-pressure changes between the aquifers and aquitards, can be seen at this site in the Antelope Valley, Mojave Desert, California.

The responses to changing water levels following eight decades of ground-water development suggest that stresses directly driving much of the compaction are somewhat insulated from the changing stresses caused by short-term water-level variations in the aquifers.

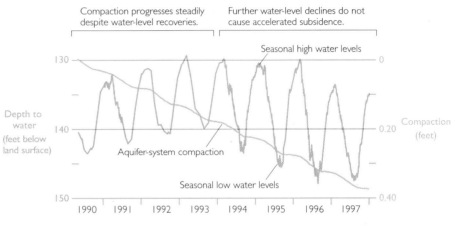

(Michelle Sneed, USGS, written communication, 1998)

SANTA CLARA VALLEY, CALIFORNIA

A case of arrested subsidence

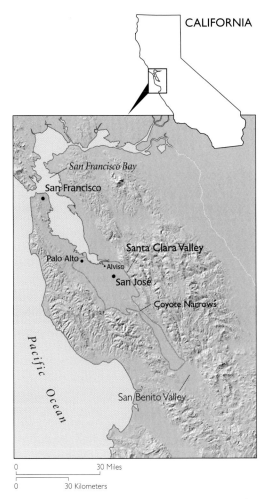

CALIFORNIA

The Santa Clara Valley is part of a structural trough that extends about 90 miles southeast from San Francisco. The northern third of the trough is occupied by the San Francisco Bay, the central third by the Santa Clara Valley, and the southern third by the San Benito Valley. The northern Santa Clara Valley, roughly from Palo Alto to the Coyote Narrows (10 miles southeast of downtown San Jose), is now densely populated and known as "Silicon Valley," the birthplace of the global electronics industry.

In the first half of this century, the Santa Clara Valley was intensively cultivated, mainly for fruit and vegetables. The extensive orchards, dominated by apricots, plums, cherries, and pears, led local boosters to dub the area a Garden of Eden or "The Valley of Heart's Delight." In the post-World War II era (circa 1945–1970), rapid population growth was associated with the transition from an agriculturally based economy to an industrial and urban economy. The story of land subsidence in the Santa Clara Valley is closely related to the changing land and water use and the importation of surface water to support the growing urban population.

San Jose and its surrounding communities sprawl across the Santa Clara Valley. The view is looking southeast from downtown San Jose.

S.E. Ingebritsen and David R. Jones
U.S. Geological Survey, Menlo Park, California

(Air Flight Service)

The Santa Clara Valley was a premier fruit growing region in the early part of the 20th century. The landscape was dotted with family orchards, each with its own well (note well house far right).

(Alice Iola Hare, Bancroft Library, UC Berkeley)

(George E. Hyde & Co. 1915-1921, Bancroft Library, UC Berkeley)

The Santa Clara Valley was the first area in the United States where land subsidence due to ground-water withdrawal was recognized (Tolman and Poland, 1940). It was also the first area where organized remedial action was undertaken, and subsidence was effectively halted by about 1969. The ground-water resource is still heavily used, but importation of surface water has reduced ground-water pumping and allowed an effective program of ground-water recharge that prevents ground-water levels from approaching the historic lows of the 1960s. The unusually well-coordinated and effective conjunctive use of surface water and ground water in the Santa Clara Valley is facilitated by the fact that much of the Valley is served by a single water-management agency, the Santa Clara Valley Water District.

GROUND-WATER PUMPING SUPPLIED ORCHARDS AND, EVENTUALLY, CITIES

This free-flowing artesian well was capped to prevent waste (1910).

The moderate climate of the Santa Clara Valley has distinct wet and dry periods. During the wet season (November to April), average rainfall ranges from a high of about 40 inches in the low, steep mountain ranges to the southwest to a low of about 14 inches on the valley floor—rates that are generally insufficient to support specialty crops. Early irrigation efforts depended upon local diversions of surface water, but the acreage that could be irrigated in this manner was very limited. By the 1860s, wells were in common use.

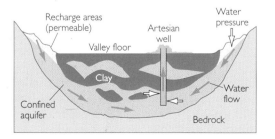

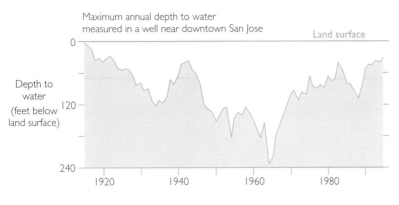

Maximum annual depth to water
measured in a well near downtown San Jose

Depth to
water
(feet below
land surface)

In the late 1800s construction of railroads, refrigerator cars, and improved canning techniques gave farmers access to the growing California and eastern markets for perishable crops. The planting of orchards and associated ground-water pumping increased rapidly into the 1900s.

In the late 1880s most wells in the area between downtown San Jose and Alviso and along the Bay northwest and northeast of Alviso were artesian. That is, water flowed freely without needing to be pumped. In fact, there was substantial waste of ground water from uncapped artesian wells. The widespread artesian conditions were due to the natural hydrogeology of the Santa Clara Valley. Water levels in the artesian wells rose above the land surface because they tapped confined aquifers that have permeable connections to higher-elevation recharge areas on the flanks of the Valley but are overlain by low-permeability clay layers.

By 1920, two-thirds of the Santa Clara Valley was irrigated, including 90 percent of the orchards, and new wells were being drilled at the rate of 1,700 per year (California History Center, 1981). By the late 1920s, about 130,000 acre-feet of ground water was pumped annually to irrigate crops and support a total population of about 100,000.

Acre-Feet

Hydrologists frequently use the term acre-feet to describe a volume of water. One acre-foot is the volume of water that will cover an area of one acre to a depth of one foot. The term is especially useful where large volumes of water are being described. One acre-foot is equivalent to 43,560 cubic feet, or about 325,829 gallons!

Ground-water levels drop

Ground water was being used faster than it could be replenished. As a result, water levels were dropping and artesian wells becoming increasingly rare. By 1930, the water level in a formerly artesian USGS monitoring well in downtown San Jose had fallen 80 feet below the land surface.

Between 1920 and 1960 an average of about 100,000 acre-feet per year of ground water was used to irrigate crops. Nonagricultural use of ground water began to increase substantially during the 1940s, and by 1960 total ground-water withdrawals approached 200,000 acre-feet per year. In 1964 the water level in the USGS monitoring well in downtown San Jose had fallen to a historic low of 235 feet below the land surface.

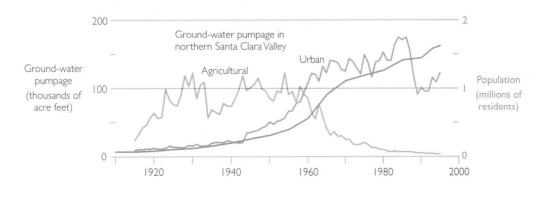

Ground-water
pumpage
(thousands of
acre feet)

Ground-water pumpage in
northern Santa Clara Valley

Agricultural

Urban

Population
(millions of
residents)

These photographs of the South Bay Yacht Club in Alviso show dramatic evidence of subsidence.

1914—The Yacht Club (building to the right) is practically at sea level.

1978—The Yacht Club is now about 10 feet below sea level, and a high levee keeps bay water from inundating Alviso.

(Santa Clara Valley Water District)

MASSIVE GROUND-WATER WITHDRAWAL CAUSED THE GROUND TO SUBSIDE

Substantial land subsidence occurred in the northern Santa Clara Valley as a result of the massive ground-water overdrafts. Detectable subsidence of the land surface (greater than 0.1 feet) took place over much of the area. The maximum subsidence occurred in downtown San Jose, where land-surface elevations decreased from about 98 feet above sea level in 1910 to about 84 feet above sea level in 1995.

Lands adjacent to the southern end of San Francisco Bay sank from 2 to 8 feet by 1969, putting 17 square miles of dry land below the high-tide level. The southern end of the Bay is now ringed with dikes to prevent landward movement of saltwater, and flood-control levees have been built to control the bayward ends of stream channels. The stream channels must now be maintained well above the surrounding land in order to provide a gradient for flow to the Bay. In the land that has sunk below the high-tide level, local storm discharge must be captured and pumped over levees in order to prevent widespread flooding.

The fact that Santa Clara Valley was subsiding became generally known in 1933, when bench marks in San Jose that were established in 1912 were resurveyed and found to have subsided 4 feet. This finding motivated the U.S. Coast and Geodetic Survey to establish a network of bench marks tied to stable bedrock on the edges of the Valley. The bench-mark network was remeasured many times between 1934 and 1967, and forms the basis for mapping subsidence.

During the 33-year period, subsidence ranged from 2 feet under the Bay and its tideland to 8 feet in San Jose and Santa Clara.

Total land subsidence, which probably began in the 1920s and continued to 1969 or later, is likely greater than shown on this map.

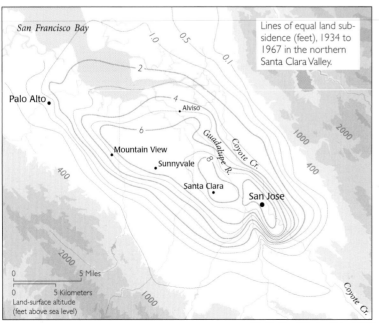

Lines of equal land subsidence (feet), 1934 to 1967 in the northern Santa Clara Valley.

(Modified from Poland and Ireland, 1988)

Subsidence had to be stopped

In 1935 and 1936, the Santa Clara Valley Water District built five storage dams on local streams to capture storm flows. This permitted controlled releases to increase ground-water recharge through streambeds. Wet years in the early 1940s enhanced both natural and artificial recharge. Although subsidence was briefly arrested during World War II, these measures proved inadequate to halt water-level declines over the long term, and, between 1950 and 1965, subsidence resumed at an accelerated rate. In 1965, increased imports of surface water allowed the Santa Clara Valley Water District to greatly expand its program of ground-water recharge, leading to substantial recovery of ground-water levels, and there has been little additional subsidence since about 1969.

In fact, as of 1995, water levels in the USGS monitoring well in downtown San Jose were only 35 feet below land surface, the highest levels observed since the early 1920s. A series of relatively wet years in the mid-1990s even caused a return to artesian conditions in some areas near San Francisco Bay. Some capped and long-forgotten wells near the Bay began to leak and were thereby rediscovered!

Subsidence in the Santa Clara Valley was caused by the decline of artesian pressures and the resulting increase in the effective overburden load on the water-bearing sediments. The sediments compacted under the increasing stress and the land surface sank. Most of the compaction occurred in fine-grained clay deposits (aquitards), which are more compressible, though less permeable, than coarser-grained sediments. The low permeability of the clay layers retards and smooths the compaction of the aquifer system relative to the water-level variations in the permeable aquifers. Since 1969, despite water-level recoveries, a small amount of additional residual com-

Land subsidence was a result of intensive ground-water pumping and the subsequent drop in water levels. Once pumping was stabilized by the introduction of imported surface water, subsidence was arrested.

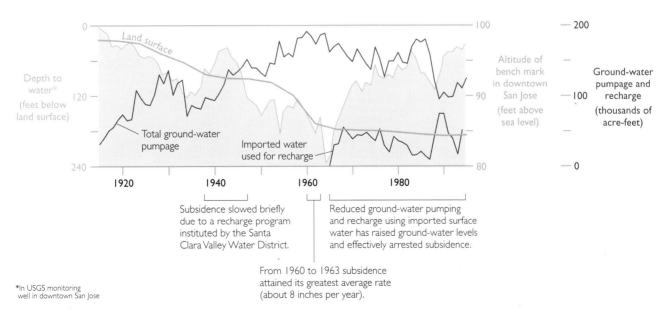

Total ground-water pumpage

Imported water used for recharge

Subsidence slowed briefly due to a recharge program instituted by the Santa Clara Valley Water District.

Reduced ground-water pumping and recharge using imported surface water has raised ground-water levels and effectively arrested subsidence.

From 1960 to 1963 subsidence attained its greatest average rate (about 8 inches per year).

*In USGS monitoring well in downtown San Jose

paction and subsidence has accrued. The total subsidence has been large and chiefly permanent, but future subsidence can be controlled if ground-water levels are maintained safely above their subsidence thresholds.

Surface water is delivered for use in the Valley

To balance Santa Clara Valley's water-use deficit, surface water has been imported from northern and eastern California via aqueducts—Hetch Hetchy (San Francisco Water Department, 1951-), the California State Water Project (1965-), and the Federal San Felipe Water Project (1987-). Much of the imported water also feeds into various local distribution lines. But presently about one-fourth of the water imported by the Santa Clara Valley Water District (about 40,000 of the 150,000 acre-feet total) is used for ground-water recharge.

The aquifer systems are used for natural storage and conveyance, in preference to constructing expensive surface-storage and conveyance systems. In order to avoid recurrence of the land subsidence that plagued the Valley prior to 1969, ground-water levels are maintained well above their historic lows, even during drought periods. For example, ground-water levels beneath downtown San Jose were maintained even during the major California droughts of 1976–77 and 1987–91. In order to avoid large ground-water overdrafts, the Water District aggressively encourages water conservation during drought periods. Per-capita water use under current conditions is much lower than in the agrarian past. Today, about 350,000 acre-feet of surface and ground water meet the annual requirements of a countywide population of about 1,600,000, and per-capita water use is only about one-fifth of the 1920 level.

The economic impact can only be approximated

The direct costs of land subsidence in the Santa Clara Valley include the cost of constructing levees around the southern end of San Francisco Bay and the bayward ends of stream channels, main-

The South Bay aqueduct conveys water from the Sacramento-San Joaquin Delta to the Santa Clara Valley.

(Santa Clara Valley Water District)

Water imports allow water managers to raise ground-water levels by reducing net ground-water extraction.

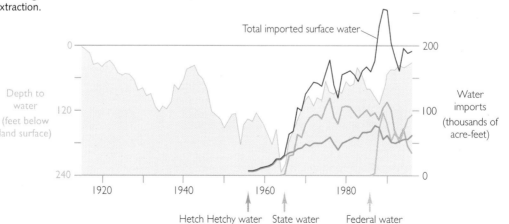

Santa Clara Valley Water District
Ground-water recharge system

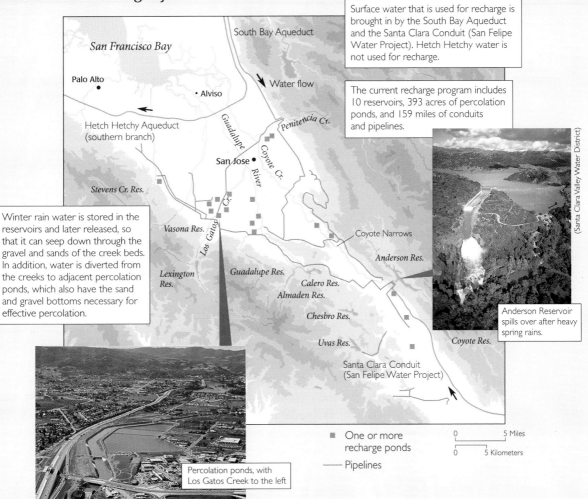

Surface water that is used for recharge is brought in by the South Bay Aqueduct and the Santa Clara Conduit (San Felipe Water Project). Hetch Hetchy water is not used for recharge.

The current recharge program includes 10 reservoirs, 393 acres of percolation ponds, and 159 miles of conduits and pipelines.

Winter rain water is stored in the reservoirs and later released, so that it can seep down through the gravel and sands of the creek beds. In addition, water is diverted from the creeks to adjacent percolation ponds, which also have the sand and gravel bottoms necessary for effective percolation.

San Francisco Bay

Palo Alto

Alviso

Hetch Hetchy Aqueduct (southern branch)

Water flow

Guadalupe

Penitencia Cr.

Coyote Cr.

San Jose

Coyote Cr. River

Stevens Cr. Res.

Los Gatos Cr.

Vasona Res.

Coyote Narrows

Anderson Res.

Lexington Res.

Guadalupe Res.

Calero Res.

Almaden Res.

Chesbro Res.

Uvas Res.

Coyote Res.

Santa Clara Conduit (San Felipe Water Project)

(Santa Clara Valley Water District)

Anderson Reservoir spills over after heavy spring rains.

■ One or more recharge ponds

— Pipelines

0   5 Miles
0   5 Kilometers

Percolation ponds, with Los Gatos Creek to the left

(Santa Clara Valley Water District)

NATURAL CONDITIONS

Conditions are favorable for recharge in the upper reaches of several streams because there is an abundance of coarse sand and gravel deposits and the aquifer system is generally unconfined; that is, fluid pressure in the aquifer is not confined by any overlying lenses of low-permeability clay. Nearer to the Bay, sediments tend to be finer-grained, and the exploited ground-water system is generally confined by low-permeability materials that impede recharge.

RECHARGE FACILITIES

The first percolation facilities in the Santa Clara Valley were built in the 1930s. They relied on capturing local surface runoff, and proved inadequate to keep pace with the rate of ground-water extraction. The volume of artificial recharge was increased significantly when additional imported surface water became available in 1965. Artificial recharge rates in the 1970s were sufficient to reverse ground-water level declines and arrest subsidence.

COST-BENEFIT

In 1984, a cost-benefit approach was used to estimate the value of artificial ground-water recharge in the Santa Clara Valley (Reichard and Bredehoeft, 1984). The benefits of reduced ground-water pumping costs and reduced subsidence were found to be greater than the total costs of continuing the artificial recharge program. A second analysis compared the costs of artificial recharge with the cost of a surface system that would achieve the same storage and conveyance of water. The costs of artificial recharge proved to be much less than the costs of an equivalent surface system.

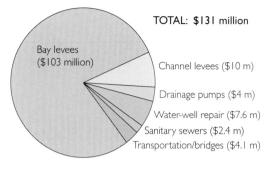

TOTAL: $131 million

Bay levees ($103 million)

Channel levees ($10 m)

Drainage pumps ($4 m)

Water-well repair ($7.6 m)

Sanitary sewers ($2.4 m)

Transportation/bridges ($4.1 m)

Direct costs of land subsidence in the Santa Clara Valley in 1979 dollars.

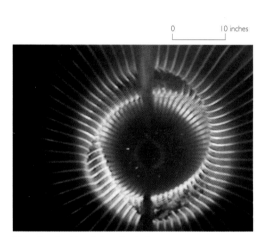

0   10 inches

This view looking into a typical collapsed well screen shows the damage caused by compaction. This photograph was made by lowering a light into the well, followed by a camera; the crumpled vertical ribbing of the steel well screen produced this radiating effect.

taining salt-pond levees, raising grades for railroads and roads, enlarging or replacing bridges, enlarging sewers and adding sewage pumping stations, and constructing and operating storm-drainage pumping stations in areas that have subsided below the high-tide level. Most of these direct costs were incurred during the era of active subsidence. In 1981 Lloyd C. Fowler, former Chief Engineer of the Santa Clara Valley Water District, estimated the direct costs of subsidence to be $131,100,000 in 1979 dollars, a figure that translates to about $300,000,000 in 1998 dollars. The ongoing cost of maintaining levees and pumping facilities can also be attributed mainly to subsidence. In fact, as of this writing, the U.S. Army Corps of Engineers is building a substantial system for flood control along the lower Guadalupe River channel, with design requirements (and associated expense) influenced by past subsidence.

Some of Fowler's estimates of direct costs deserve further explanation. Land subsidence was estimated to have damaged or destroyed about 1,000 wells in the 5-year period 1960 to 1965, and the cost estimate was based on the cost of repair. By the 1960s most large wells in the Santa Clara Valley extended to depths of 400 feet or more. Many well casings were buckled or collapsed by the compaction of clay lenses at depths more than 200 feet below the land surface. The compacting clay caused the casing to buckle and eventually collapse. The cost estimate cited for the Bay levees as of 1979 applies only to the publicly maintained flood-protection levees, and likely underestimates the total cost. An additional, unknown cost was incurred by a salt company that maintained levees on 30 square miles of salt ponds within the original bayland area. Land subsidence has permanently increased the risk of saltwater flooding in case of levee breaks and the potential for saltwater intrusion of shallow aquifers.

Careful management will continue

The Santa Clara Valley Water District is currently managing the ground-water basin in a conservative fashion in order to avoid further subsidence. Their management strategy depends on continued availability of high-quality surface water from State and Federal projects that import water from massive diversion facilities in the southern part of the Sacramento-San Joaquin Delta. As we describe in another case study, these diversion facilities themselves are threatened by land subsidence within the Delta. Thus the prognosis for land subsidence in the Santa Clara Valley depends in part on subsidence rates and patterns in the Delta. Because much of California relies on large-scale interbasin water transfers, subsidence and water-quality issues in many parts of the State are complexly interrelated.

SAN JOAQUIN VALLEY, CALIFORNIA

Largest human alteration of the Earth's surface

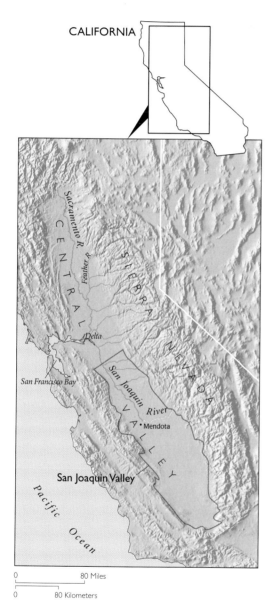

CALIFORNIA

Sacramento R.
Feather R.
CENTRAL
SIERRA
Delta
NEVADA
San Francisco Bay
San Joaquin River
Mendota
VALLEY
San Joaquin Valley
Pacific
Ocean

0   80 Miles

0   80 Kilometers

Mining ground water for agriculture has enabled the San Joaquin Valley of California to become one of the world's most productive agricultural regions, while simultaneously contributing to one of the single largest alterations of the land surface attributed to humankind. Today the San Joaquin Valley is the backbone of California's modern and highly technological agricultural industry. California ranks as the largest agricultural producing state in the nation, producing 11 percent of the total U.S. agricultural value. The Central Valley of California, which includes the San Joaquin Valley, the Sacramento Valley, and the Sacramento-San Joaquin Delta, produces about 25 percent of the nation's table food on only 1 percent of the country's farmland (Cone, 1997).

In 1970, when the last comprehensive surveys of land subsidence were made, subsidence in excess of 1 foot had affected more than 5,200 square miles of irrigable land—one-half the entire San Joaquin Valley (Poland and others, 1975). The maximum subsidence, near Mendota, was more than 28 feet.

1925

1955

SAN JOAQUIN VALLEY
CALIFORNIA
BM S661
SUBSIDENCE 9M
1925-1977

1977

Approximate location of maximum subsidence in United States identified by research efforts of Joseph Poland (pictured). Signs on pole show approximate altitude of land surface in 1925, 1955, and 1977. The pole is near benchmark S661 in the San Joaquin Valley southwest of Mendota, California.

Devin Galloway and Francis S. Riley
U.S. Geological Survey, Menlo Park, California

Since the early 1970s land subsidence has continued in some locations, but has generally slowed due to reductions in ground-water pumpage and the accompanying recovery of ground-water levels made possible by supplemental use of surface water for irrigation. The surface water is diverted principally from the Sacramento-San Joaquin Delta and the San Joaquin, Kings, Kern and Feather Rivers. Two droughts since 1975 have caused surface-water deliveries in the valley to be sharply curtailed, and demonstrated the valley's vulnerability to continued land subsidence when ground-water pumpage is increased.

The history of land subsidence in the San Joaquin Valley is integrally linked to the development of agriculture and the availability of water for irrigation. Further agricultural development without accompanying subsidence is dependent on the continued availability of surface water, which is subject to uncertainties due to climatic variability and pending regulatory decisions.

Land subsidence in the San Joaquin Valley was first noted in 1935 when I. H. Althouse, a consulting engineer, called attention to the possibility of land subsidence near the Delano (Tulare-Wasco) area. The process was first described in print by Ingerson (1941, p. 40–42), who presented a map and profiles of land subsidence based on comparison of leveling of 1902, 1930, and 1940. Four types of subsidence are known to occur in the San Joaquin Valley. In order of decreasing magnitude they are (1) subsidence caused by aquifer-system compaction due to the lowering of ground-water levels by sustained ground-water overdraft; (2) subsidence caused by the hydrocompaction of moisture-deficient deposits above the water-table; (3) subsidence related to fluid withdrawal from oil and gas fields; and (4) subsidence related to crustal neotectonic movements. Aquifer-system compaction and hydrocompaction have significantly lowered the land surface in the valley since about the 1920s, and our review of the subsidence problems there is limited to these two primary causes.

THE SAN JOAQUIN VALLEY IS PART OF A GREAT SEDIMENT-FILLED TROUGH

The San Joaquin Valley comprises the southern two-thirds of the Central Valley of California. Situated between the towering Sierra Nevada on the east, the Diablo and Temblor Ranges to the west, and the Tehachapi Mountains to the south, the valley occupies a trough created by tectonic forces related to the collision of the Pacific and North American Plates. The trough is filled with marine sediments overlain by continental sediments, in some places thousands of feet deep, deposited largely by streams draining the mountains, and partially in lakes that inundated portions of the valley floor from time to time. More than half the thickness of the continental sediments is composed of fine-grained (clay, sandy clay, sandy silt, and silt) stream (fluvial) and lake (lacustrine) deposits susceptible to compaction.

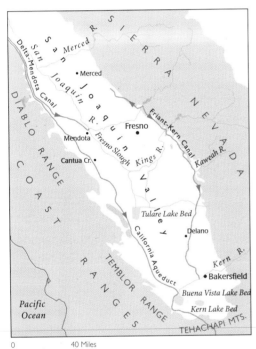

Meltwater from the Sierra snowpack recharges ground water in the San Joaquin Valley and supplies surface water during the dry summer months.

(California Department of Water Resources)

The valley floor, comprising about 10,000 square miles, is arid to semiarid, receiving an average of 5 to 16 inches of rainfall annually. Most of the streamflow in the valley enters from the east side in streams draining the western Sierra Nevada, where much of the precipitation occurs as snow. The San Joaquin River begins high in the Sierra Nevada and descends onto the valley floor, where it takes a northerly flow path toward the Sacramento-San Joaquin Delta. On its course northward to the Delta it collects streamflow from the central and northern portions of the valley. The southern valley receives streamflow from the Kings, Kaweah, and Kern Rivers, which issue from steeply plunging canyons onto broad, extensive alluvial fans. Over many thousands of years, the natural flow of these rivers distributed networks of streams and washes on the slopes of the alluvial fans and terminated in topographically closed sinks, such as Tulare Lake, Kern Lake, and Buena Vista Lake. Streams draining the drier western slopes and Coast Ranges adjacent to the valley are intermittent or ephemeral, flowing only episodically. Precipitation and streamflow in the valley vary greatly from year to year.

Pumping for irrigation altered the ground-water budget

Ground water occurs in shallow, unconfined (water table) or partially-confined aquifers throughout the valley. Such aquifers are particularly important near the margins of the valley and near the toes of younger alluvial fans. A laterally extensive lacustrine clay known as the Corcoran Clay is distributed throughout the central and western valley. The Corcoran Clay, which varies in thickness from a feather edge to about 160 feet beneath the present bed of Tulare Lake, confines a deeper aquifer system that comprises fine-grained aquitards interbedded with coarser aquifers. Most of the subsidence measured in the valley has been correlated with the distribution of ground-water pumpage and the reduction of water levels in the deep confined aquifer system.

Under natural conditions before development, ground water in the alluvial sediments was replenished primarily by infiltration through stream channels near the valley margins. The eastern-valley streams carrying runoff from the Sierra Nevada provided most of the recharge for valley aquifers. Some recharge also occurred from precipitation falling directly on the valley floor and from stream and lake seepage occurring there. Over the long term, natural replenishment was dynamically balanced by natural depletion through ground-water discharge, which occurred primarily through evapotranspiration and contributions to streams flowing into the Delta. The areas of natural discharge in the valley generally corresponded with the areas of flowing, artesian wells mapped in an early USGS investigation (Mendenhall and others, 1916). Direct ground-water outflow to the Delta is thought to have been negligible.

Today, nearly 150 years since water was first diverted at Peoples Weir on the Kings River and more than 120 years after the first irrigation colonies were established in the valley, intensive development of ground-water resources for agricultural uses has drastically altered the valley's water budget. The natural replenishment of the aquifer systems has remained about the same, but more water has discharged than recharged the aquifer system; the deficit may have amounted to as much as 800,000 acre-feet per year during the late 1960s (Williamson et al., 1989). Most of the surface water now being imported is transpired by crops or evaporated from the soil. The amount of surface-water outflow from the valley has actually been

PREDEVELOPMENT

Ground water flowed from the mountains toward the center of the valley where it discharged into streams or through evapotranspiration.

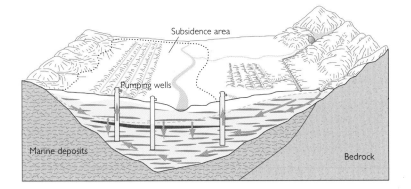

POSTDEVELOPMENT

Ground water flows generally downward and toward pumping centers.

By pumping the vast reserves of ground water, farmers have developed the San Joaquin Valley into a major agricultural region.

(California Department of Water Resources)

reduced compared to predevelopment conditions. Ground water in the San Joaquin Valley has generally been depleted and redistributed from the deeper aquifer system to the shallow aquifer system. This has created problems of ground-water quality and drainage in the shallow aquifer system, which is infiltrated by excess irrigation water that has been exposed to agricultural chemicals and natural salts concentrated by evapotranspiration.

A STABLE WATER SUPPLY IS DEVELOPED FOR IRRIGATION

In the San Joaquin Valley, irrigated agriculture surged after the 1849 Gold Rush and again in 1857, when the California Legislature passed an act that promoted the drainage and reclamation of river-bottom lands (Manning, 1967). By 1900, much of the flow of the Kern River and the entire flow of the Kings River had been diverted through canals and ditches to irrigate lands throughout the southern part of the valley (Nady and Laragueta, 1983). Because no significant storage facilities accompanied these earliest diversions, the agricultural water supply, and hence crop demand, was largely limited by the summer low-flows. The restrictions imposed by the need for constant surface-water flows, coupled with a drought occurring around 1880 and the fact that, by 1910, nearly all the available surface-water supply in the San Joaquin Valley had been diverted, prompted the development of ground-water resources.

The first development of the ground-water resource occurred in regions where shallow ground water was plentiful, and particularly where flowing wells were commonplace, near the central part of the valley around the old lake basins. Eventually, the yields of flowing wells diminished as water levels were reduced, and it became necessary to install pumps in wells to sustain flow rates. Around 1930, the development of an improved deep-well turbine pump and rural electrification enabled additional ground-water development for irrigation. The ground-water resource had been established as a reliable, stable water-supply for irrigation. Similar histories were repeated in many other basins in California and throughout the Southwest, where surface water was limited and ground water was readily available.

Overhead and flood irrigation supply water to a wide range of crops.

(California Department of Water Resources)

WATER WITHDRAWAL CAUSED LAND SUBSIDENCE

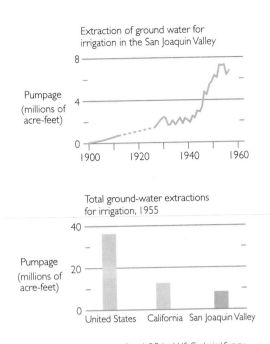

Extraction of ground water for irrigation in the San Joaquin Valley

Total ground-water extractions for irrigation, 1955

(Joseph F. Poland, U.S. Geological Survey, written communication, ca 1957)

Shortly after the completion of the Delta-Mendota Canal by the U.S. Bureau of Reclamation in 1951, subsidence caused by withdrawal of ground water in the northern San Joaquin Valley had begun to raise concerns, largely because of the impending threat to the canal and the specter of remedial repairs. Because of this threat to the canal, and in order to help plan other major canals and engineering proposed for construction in the subsiding areas, the USGS, in cooperation with the California Department of Water Resources, began an intensive investigation into land subsidence in the San Joaquin Valley. The objectives were to determine the causes, rates, and extent of land subsidence and to develop scientific criteria for the estimation and control of subsidence. The USGS concurrently began a federally funded research project to determine the physical principles and mechanisms governing the expansion and compaction of aquifer systems resulting from changes in aquifer hydraulic heads. Much of the material presented here is drawn from these studies.

In 1955, about one-fourth (almost 8 million acre-feet) of the total ground water extracted for irrigation in the United States was pumped in the San Joaquin Valley. The maximum changes in water levels occurred in the western and southern portions of the valley, in the deep confined aquifer system. More than 400 feet of water-

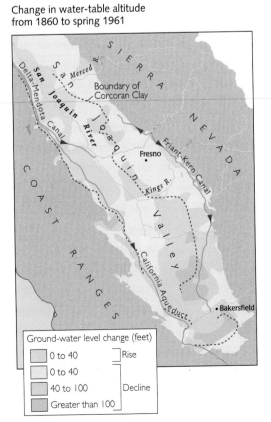

Change in water-table altitude from 1860 to spring 1961

Ground-water level change (feet)

0 to 40	Rise
0 to 40	
40 to 100	Decline
Greater than 100	

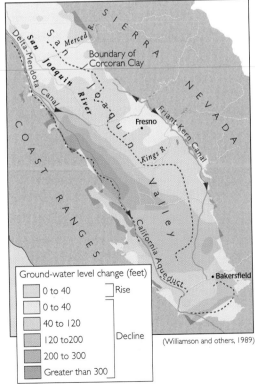

Change in water level in the deep confined aquifer system from 1860 to spring 1961

Ground-water level change (feet)

0 to 40	Rise
0 to 40	
40 to 120	
120 to 200	Decline
200 to 300	
Greater than 300	

(Williamson and others, 1989)

By 1971 the growing use of imported surface-water supplies surpasses the use of local ground-water supplies, but the effects of drought reverse this trend in 1977.

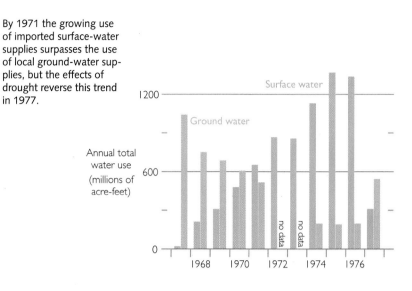

Annual total water use (millions of acre-feet)

level decline occurred in some west-side areas in the deep aquifer system. Until 1968, irrigation water in these areas was supplied almost entirely by ground water. As of 1960, water levels in the deep aquifer system were declining at a rate of about 10 feet per year. Western and southern portions of the valley generally experienced more than 100 feet of water-level decline in the deep aquifer system. Water levels in the southeastern and eastern portions of the valley were generally less affected because some surface water was also available for irrigation. In the water-table aquifer, few areas exceeded 100 feet of water-level decline, but a large portion of the southern valley did experience declines of more than 40 feet. In some areas on the northwest side, the water-table aquifer rose up to 40 feet due to infiltration of excess irrigation water.

Accelerated ground-water pumpage and water-level declines, principally in the deep aquifer system during the 1950s and 1960s, caused about 75 percent of the total volume of land subsidence in the San Joaquin Valley. By the late 1960s, surface water was being diverted to agricultural interests from the Sacramento-San Joaquin Delta and the San Joaquin River through federal reclamation projects and from the Delta through the newly completed, massive State (California) Water Project. Less-expensive water from the Delta-Mendota Canal, the Friant-Kern Canal, and the California Aqueduct largely supplanted ground water for crop irrigation. Ground-water levels began a dramatic period of recovery, and subsidence slowed or was arrested over a large part of the affected area. Water levels in the deep aquifer system recovered as much as 200 feet in the 6 years from 1967 to 1974 (Ireland and others, 1984).

When water levels began to recover in the deep aquifer system, aquifer-system compaction and land subsidence began to abate, although many areas continued to subside, albeit at a lesser rate. During the period from 1968 to 1974, water levels measured in an observation well near Cantua Creek recovered more than 200 feet while another 2 feet of subsidence continued to accrue. This apparent contradiction is the result of the time delay in the compaction

Land subsidence from 1926 to 1970

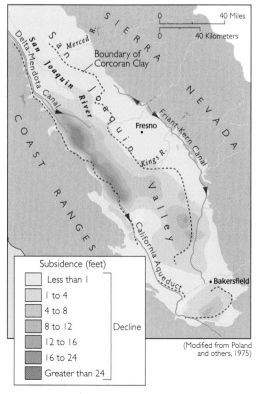

Subsidence (feet)
- Less than 1
- 1 to 4
- 4 to 8
- 8 to 12
- 12 to 16
- 16 to 24
- Greater than 24

Decline

(Modified from Poland and others, 1975)

To supplement local ground-water supplies, the California Aqueduct (left) conveys water from the Delta to the dry southern valleys.

(California Department of Water Resources)

of the aquitards in the aquifer system. The delay is caused by the time that it takes for pore-fluid pressures in the aquitards to equilibrate with the pressure changes occurring in the aquifers, which are much more responsive to the current volume of ground-water being pumped (or not pumped) from the aquifer system. The time needed for pressure equilibration depends largely on the thickness and permeability of the aquitards. Typically, as in the San Joaquin Valley, centuries will be required for most of the pressure equilibration to occur, and therefore for the ultimate compaction to be realized. Swanson (1998) states that "Subsidence is continuing in all historical subsidence areas..., but at lower rates than before...."

Since 1974, land subsidence has been greatly slowed or largely arrested but remains poised to resume. In fact, during the severe

When water levels recover, compaction and land subsidence can abate.

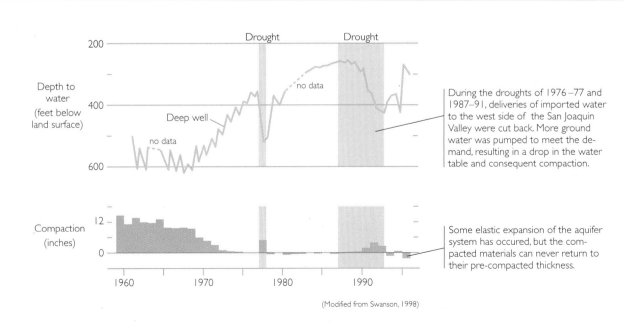

During the droughts of 1976–77 and 1987–91, deliveries of imported water to the west side of the San Joaquin Valley were cut back. More ground water was pumped to meet the demand, resulting in a drop in the water table and consequent compaction.

Some elastic expansion of the aquifer system has occured, but the compacted materials can never return to their pre-compacted thickness.

(Modified from Swanson, 1998)

In the major subsiding areas, subsidence has continued except for a slight leveling off in the mid 1970s.

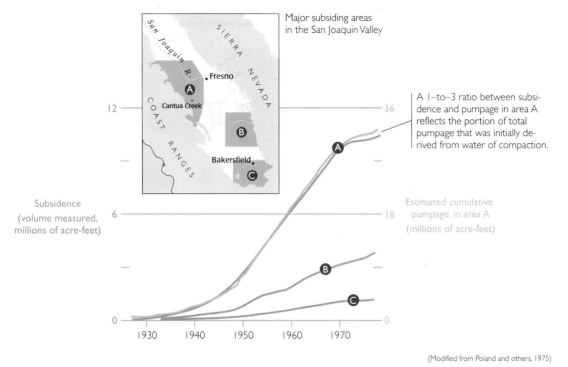

Major subsiding areas in the San Joaquin Valley

A 1–to–3 ratio between subsidence and pumpage in area A reflects the portion of total pumpage that was initially derived from water of compaction.

Subsidence (volume measured, millions of acre-feet)

Estimated cumulative pumpage, in area A (millions of acre-feet)

(Modified from Poland and others, 1975)

droughts in California in 1976–77 and 1987–91, diminished deliveries of imported water prompted some water agencies and farmers, especially in the western valley, to refurbish old pumping plants, drill new wells, and begin pumping ground water to make up for cutbacks in the imported water supply. The decisions to renew ground-water pumpage were encouraged by the fact that ground-water levels had recovered nearly to predevelopment levels. During the 1976–77 drought, after only one-third of the peak annual pumpage volumes of the 1960s had been produced, ground-water levels rapidly declined more than 150 feet over a large area and subsidence resumed. Nearly 0.5 feet of subsidence was measured in 1977 near Cantua Creek. This scenario was repeated during the more recent 1987–91 drought. It underscores the sensitive dependence between subsidence and the dynamic state of imported-water availability and use.

That a relatively small amount of renewed pumpage caused such a rapid decline in water levels reflects the reduced ground-water storage capacity—lost pore space—caused by aquifer-system compaction. It demonstrates the nonrenewable nature of the resource embodied in the "water of compaction." It emphasizes the fact that extraction of this resource, available only on the first cycle of large-scale drawdown, must be viewed, like more traditional forms of mining, in terms not only of its obvious economic return but also its less readily identifiable costs.

Hydrocompaction
Compaction near the surface

Hydrocompaction produces an undulating surface of hollows and hummocks with local relief, typically of 3 to 5 feet. In this view of a furrowed field, the hollows are filled with irrigation water.

Hydrocompaction—compaction due to wetting— is a near-surface phenomenon that produces land-surface subsidence through a mechanism entirely different from the compaction of deep, overpumped aquifer systems. Both of these processes accompanied the expansion of irrigated agriculture onto the arid, gentle slopes of the alluvial fans along the west side and south end of the San Joaquin Valley. Initially, the distinction between them, and their relative contributions to the overall subsidence problem, were not fully recognized.

In the 1940s and 50s farmers bringing virgin valley soils under cultivation found that standard techniques of flood irrigation caused an irregular settling of their carefully graded fields, producing an undulating surface of hollows and hummocks with local relief, typically of 3 to 5 feet. Where water flowed or ponded continuously for months, very localized settlements of 10 feet or more might occur on susceptible soils. These consequences of artificial wetting seriously disrupted the distribution of irrigation water and damaged pipelines, power lines, roadways, airfields, and buildings. In contrast to the broad, slowly progressive and generally smooth subsidence due to deep-seated aquifer-system compaction, the irregular, localized, and often rapid differential subsidence due to hydrocompaction was readily discernible without instrumental surveys. Recognition of its obvious impact on the design and construction of the proposed California Aqueduct played a major role in the initiation in 1956 of intensive studies to identify, characterize, and quantify the subsidence processes at work beneath the surface of the San Joaquin Valley.

MECHANISMS OF COMPACTION WERE ANALYZED

The mechanisms and requisite conditions for hydrocompaction, initially known as "near-surface subsidence," were investigated by means of laboratory tests on soil cores from depths to 100 or more feet, and by continuously flooded test plots equipped with subsurface benchmarks at various depths and, in some cases, with soil-moisture probes.

The combined field and laboratory studies demonstrated that hydrocompaction occurred only in alluvial-fan sediments above the highest prehistoric water table and in areas where sparse rainfall and ephemeral runoff had never

Hydrocompaction caused surface cracks and land subsidence at experimental Test Plot B, Fresno County.

Mudflow containing hydrocompactible sediments, western Fresno County (1961)

penetrated below the zone subject to summer desiccation by evaporation and transpiration. Under these circumstances the initial high porosity of the sediments (often enhanced by numerous bubble cavities and desiccation cracks) is sun-baked into the deposits and preserved by their high dry strength, even as they are subjected to the increasing load of 100 or more feet of accumulating overburden. In the San Joaquin Valley, such conditions are associated with areas of very low average rainfall and infrequent, flashy, sediment-laden runoff from small, relatively steep upland watersheds that are underlain by easily erodable shales and mudstones. The resulting muddy debris flows and poorly sorted stream sediments typically contain montmorillonite clay in proportions that cause it to act, when dry, as a strong interparticulate bonding agent. When water is first applied in quantities sufficient to penetrate below the root zone the clay bonds are drastically weakened by wetting, and the weight of the overburden crushes out the excess porosity. The process of densifying to achieve the strength required to support the existing overburden may reduce the bulk volume by as much as 10 percent, the amounts increasing with increasing depth and overburden load.

Most of the potential hydrocompaction latent in anomalously dry, low-density sediments is realized as rapidly as the sediments are thoroughly wetted. Thus the progression of a hydrocompaction event is controlled largely by the rate at which the wetting front of percolating water can move downward through the sediments. A site underlain by a thick sequence of poorly permeable sediments may continue to subside for months or years as the slowly descending wetting front weakens progressively deeper deposits. If the surface water source is seasonal or intermittent, the progression is further delayed.

Localized compaction beneath a water-filled pond or ditch often leads to vertical shear failure at depth between the water-weakened sediments and the surrounding dry material. At the surface this process surrounds the subsiding flooded area with an expanding series of concentric tensional fissures having considerable vertical offset—a severely destructive event when it occurs beneath an engineered structure.

The hazards presented by hydrocompaction are somewhat mitigated by the fact that the process goes rapidly to completion with the initial thorough wetting, and is not subject to reactivation through subsequent cycles of decreasing and increasing moisture content. However, if the volume of water that infiltrates the surface on the first wetting cycle is insufficient to wet the full thickness of susceptible deposits, then the process will propagate to greater depths on subsequent applications, resulting in renewed subsidence. Also, an increase in the surface load such as a bridge footing or a canal full of water can cause additional compaction in prewetted sediments.

Studies undertaken in the mid-1950s led to a better understanding of hydrocompaction and to the identification of long reaches of the California Aqueduct route that were underlain by deposits susceptible to hydrocompaction. Construction of the aqueduct through these reaches was preceded by prewetting, and thus compacting to a nearly stable state, the full thickness of susceptible deposits beneath the aqueduct alignment. These measures added more than two years and tens of millions of dollars to the cost of the project.

Prewetting a new section of the California Aqueduct to precompact shallow deposits susceptible to hydrocompaction (near toe of Moreno Gulch, 1963)

MANY COSTS OF LAND SUBSIDENCE ARE HIDDEN

The economic impacts of land subsidence in the San Joaquin Valley are not well known. Damages directly related to subsidence have been identified, and some have been quantified. Other damages indirectly related to subsidence, such as flooding and long-term environmental effects, merit additional assessment. Some of the direct damages have included decreased storage in aquifers, partial or complete submergence of canals and associated bridges and pipe crossings, collapse of well casings, and disruption of collector drains and irrigation ditches. Costs associated with these damages have been conservatively estimated at $25,000,000 (EDAW-ESA, 1978). These estimates are not adjusted for changing valuation of the dollar, and do not fully account for the underreported costs associated with well rehabilitation and replacement. When the costs of lost property value due to condemnation, regrading irrigated land, and replacement of irrigation pipelines and wells in subsiding areas are included, the annual costs of subsidence in the San Joaquin Valley soar to $180 million per year in 1993 dollars (G. Bertoldi and S. Leake, USGS, written communication, March 30, 1993).

HOUSTON-GALVESTON, TEXAS

Managing coastal subsidence

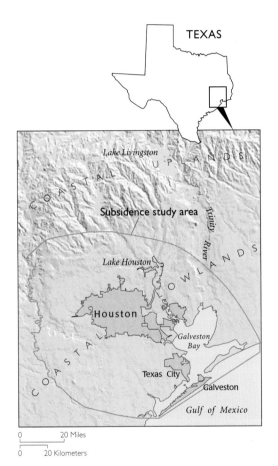

TEXAS

Lake Livingston

COASTAL UPLANDS

Subsidence study area

Lake Houston

Trinity River

COASTAL LOWLANDS

Houston

Galveston Bay

COASTAL

Texas City

Galveston

Gulf of Mexico

0   20 Miles

0   20 Kilometers

The greater Houston area, possibly more than any other metropolitan area in the United States, has been adversely affected by land subsidence. Extensive subsidence, caused mainly by ground-water pumping but also by oil and gas extraction, has increased the frequency of flooding, caused extensive damage to industrial and transportation infrastructure, motivated major investments in levees, reservoirs, and surface-water distribution facilities, and caused substantial loss of wetland habitat.

Although regional land subsidence is often subtle and difficult to detect, there are localities in and near Houston where the effects are quite evident. In this low-lying coastal environment, as much as 10 feet of subsidence has shifted the position of the coastline and changed the distribution of wetlands and aquatic vegetation. In fact, the San Jacinto Battleground State Historical Park, site of the battle that won Texas independence, is now partly submerged. This park, about 20 miles east of downtown Houston on the shores of Galveston Bay, commemorates the April 21, 1836, victory of Texans led by Sam Houston over Mexican forces led by Santa Ana. About 100 acres of the park are now under water due to subsidence, and

A road (below right) that provided access to the San Jacinto Monument was closed due to flooding caused by subsidence.

Laura S. Coplin
U.S. Geological Survey, Houston, Texas

Devin Galloway
U.S. Geological Survey, Menlo Park, California

part of the remaining area must now be protected from the Bay by dikes that trap local rain water, which must then be removed by pumps. At many localities in the Houston area, ground-water pumpage and subsidence have also induced fault movement, leading to visible fracturing, surface offsets, and associated property damage.

Growing awareness of subsidence-related problems on the part of community and business leaders prompted the 1975 Texas legislature to create the Harris-Galveston Coastal Subsidence District, "… for the purpose of ending subsidence which contributes to, or precipitates, flooding, inundation, and overflow of any area within the District …." This unique District was authorized to issue (or refuse) well permits, promote water conservation and education, and promote conversion from ground-water to surface-water supplies. It has largely succeeded in its primary objective of arresting subsidence in the coastal plain east of Houston. However, subsidence has accelerated in fast-growing inland areas north and west of Houston, which still rely on ground water and, partly as a result, the Fort Bend Subsidence District was created by the legislature in 1989.

THE FLAT, HUMID GULF COAST IS PRONE TO FLOODING

The Houston-Galveston Bay area includes a large bay-estuary-lagoon system consisting of the Trinity, Galveston, East, and West Bays, which are separated from the Gulf of Mexico by Pelican Island, Galveston Island, and the Bolivar Peninsula. Tidal exchange occurs between the Gulf and bay system through the barrier-island and peninsula complex.

The Houston climate is subtropical; temperatures range from 45° to 93° Fahrenheit and on average about 47 inches of rain falls each year. The humid coastal plain slopes gently towards the Gulf at a rate of about 1 foot per mile. Two major rivers, the Trinity and San Jacinto, and many smaller ones traverse the plain before discharging into estuarine areas of the bay system. Another large river, the Brazos, crosses the Fort Bend Subsidence District and discharges directly into Galveston Bay. The same warm waters of the Gulf of Mexico that attract recreational and commercial fishermen, and other aquatic enthusiasts, are conducive to hurricanes and tropical storms. The Texas coast is subject to a hurricane or tropical storm about once every 2 years (McGowen and others, 1977). Storm tides associated with hurricanes have reached nearly 15 feet in Galveston . The flat-lying region is particularly prone to flooding from both riverine and coastal sources, and the rivers, their reservoirs, and

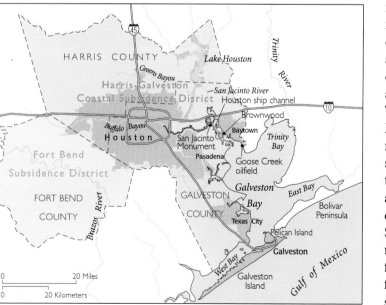

Galveston Bay near
Goose Creek

an extensive system of bayous and manmade canals are managed as part of an extensive flood-control system.

Land subsidence contributes to flooding

Land subsidence in the Houston-Galveston area has increased the frequency and severity of flooding. Near the coast, the net result of land subsidence is an apparent increase in sea level, or a relative sea-level rise: the net effect of global sea-level rise and regional land subsidence in the coastal zone. The sea level is in fact rising due to regional and global processes, both natural and human-induced. The combined effects of the actual sea-level rise and natural consolidation of the sediments along the Texas Gulf coast yield a relative sea-level rise from natural causes that locally may exceed 0.08 inches per year (Paine, 1993). Global warming is contributing to the present-day sea level rise and is expected to result in a sea-level increase of nearly 4 inches by the year 2050 (Titus and Narayanan, 1995). However, during the 20th century human-induced subsidence has been by far the dominant cause of relative sea-level rise along the Texas Gulf Coast, exceeding 1 inch per year throughout much of the affected area. This subsidence has resulted principally from extraction of ground water, and to a lesser extent oil and gas, from subsurface reservoirs. Subsidence caused by oil and gas production is largely restricted to the field of production, as contrasted to the regional-scale subsidence typically caused by ground-water pumpage.

HOUSTON'S GROWTH WAS BASED ON OIL AND GAS INDUSTRIES

Since 1897, when the population was about 25,000, the Houston area has experienced rapid growth, spurred on by the discovery of oil and establishment of the Port of Houston. In 1907 the first successful oil well was drilled, marking the beginning of the petrochemical industry that provided the economic base on which the Houston area was built and still stands. In 1925 Houston became a deep-water port when the U.S. Army Corps of Engineers completed dredging the Houston Ship Channel across Galveston Bay, up the lower reaches of the San Jacinto River, and along Buffalo Bayou to Hous-

Homes at Greens Bayou were flooded during a storm in June 1989.

(Harris-Galveston Coastal Subsidence District)

Houston (downtown can be seen top center) owes much of its development to the Houston ship channel, which is flanked by petrochemical industries and shipping facilities.

(Harris-Galveston Coastal Subsidence District)

ton. Easy access to the Gulf via the ship channel, and the discovery of additional oilfields, triggered major industrial development along the ship channel in Baytown-La Porte, Pasadena, Texas City, and Houston. The region and industry have continued to grow, and the Houston-Galveston area currently has a population of about 3 million people that is projected to grow to 4.5 million by the year 2010. Nearly half of all U.S. petrochemical production occurs in the greater Houston area. The Port of Houston is the second largest port (by tonnage shipment) in the nation, eighth largest in the world, and handles more commodities for Mexico than all Mexican ports combined. Subsidence to the east of Houston has recently been arrested by substituting imported surface water supplies for much of the ground-water pumpage, but fast growing areas to the west and north, which still depend largely on ground water, are actively subsiding.

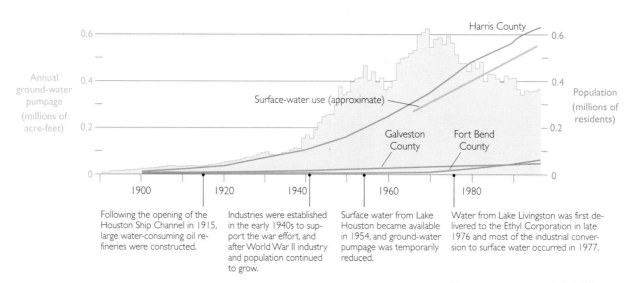

Following the opening of the Houston Ship Channel in 1915, large water-consuming oil refineries were constructed.

Industries were established in the early 1940s to support the war effort, and after World War II industry and population continued to grow.

Surface water from Lake Houston became available in 1954, and ground-water pumpage was temporarily reduced.

Water from Lake Livingston was first delivered to the Ethyl Corporation in late 1976 and most of the industrial conversion to surface water occurred in 1977.

(Compiled from Jorgenson, 1961; Gabrysch, 1987; and Houston-Galveston Coastal Subsidence District, 1996)

Goose Creek oil field
Prolific oil production produced the region's first major subsidence

Most subsidence in the Houston area has been caused by ground-water withdrawal, but the earliest subsidence was caused by oil production. In fact, the subsidence of the Goose Creek oil field on Galveston (San Jacinto) Bay was the first subsidence attributed to subsurface-fluid withdrawal to be described in the scientific literature. A dispute over the legal status of the land submerged by subsidence caused Texas courts to formally recognize the process.

"In 1917 a prolific oil field was developed near the mouth of Goose Creek, and during 1918 and subsequent years, millions of barrels of oil were removed from beneath its surface. Beginning in 1918 it became apparent that the Gaillard Peninsula, near the center of the field, and other nearby low land was becoming submerged. Elevated plank roadways or walks were built from the mainland to the derricks. Derrick floors had to be raised. Vegetation was flooded and killed, and finally all of the peninsula disappeared beneath the water… The maximum measured subsidence is now more than 3 feet and the area affected is 2½ miles long by 1½ miles wide… Outside this area no change in elevation can be detected…."

—Pratt and Johnson, 1926

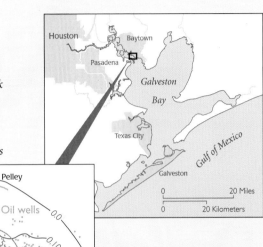

Between 1918 and 1926 subsidence was measured around Goose Creek oil-field. Lines of equal subsidence (feet) for an 8-year period are shown in grey lines—for a 1-year period, in black lines.

"There can be no doubt, …that the contours show correctly the essential fact that a local 'dishing' of the earth's surface has occurred in the Goose Creek region, the central area of greatest subsidence corresponding approximately with the center of the oil field."

—Pratt and Johnson, 1926

"Submerged land in Texas belongs to the state and only the state can grant oil and gas leases on submerged lands. Consequently, when Gaillard Peninsula became submerged, the state claimed title to it and sought not only to dispossess the fee owner and the oil and gas lessee, but also to recover from them the value of the oil and gas removed from the premises subsequent to the time when the land became submerged. The question was taken into court and finally a decision was rendered in favor of the defendants, that is, the claim of the state of Texas was denied, and the present owners continue in possession. The basis for the decision was the court's acceptance that the subsidence at Goose Creek (which the defendants admitted) was caused by an act of man, namely, the removal of large volumes of oil, gas, water, and sand from beneath the surface."

— Pratt and Johnson, 1926

Pratt and Johnson (1926) also noted that the subsided volume, calculated based on the difference between current and initial topography, amounted to about 20 per cent of the produced volume of oil, gas, water, and sand.

FAULTING FOLLOWED SUBSIDENCE

"…cracks appeared in the ground running beneath houses, across streets, and through lawns and gardens…. recurrent movement along them resulted in dropping the surface of the ground on the side toward the oil field… The movements were accompanied by slight earthquakes which shook the houses, displaced dishes, spilled water, and disturbed the inhabitants generally."

—Pratt and Johnson, 1926

This photograph taken about 1926 shows a 'fault fissure' in Pelley, one-half mile north of the oil fields. To the left of the fault, the ground had dropped about 16 inches.

Subsidence trends are related to patterns of ground-water and oil-and-gas extraction

Land subsidence first occurred in the early 1900s in areas where ground water, oil, and gas were extracted and has continued throughout the 20th century due primarily to ground-water pumpage. The patterns of subsidence in the Houston area closely follow the temporal and spatial patterns of subsurface fluid extraction.

Prior to the early 1940s there was localized subsidence caused chiefly by the removal of oil and gas along with the attendant brine, ground water and sand in oilfields such as Goose Creek. Near Texas City the withdrawal of ground water for public supply and industry caused more than 1.6 feet of subsidence between 1906 and 1943. This period also marked the beginning of a slow but steady development of ground-water resources that constituted the sole water supply for industries and communities around the Ship Channel, including Houston. By 1937 ground-water levels were falling in a growing set of gradually coalescing cones of depression centered on the areas of heavy use. Until 1942, essentially all water demand in Houston was supplied by local ground water. By 1943 subsidence had begun to affect a large part of the Houston area although the amounts were generally less than 1 foot.

A period of rapid growth in the development of ground-water resources was driven by the expansion of the petrochemical industry and other allied industries in the early 1940s through the late 1970s. By the mid-1970s, 6 or more feet of subsidence had occurred throughout an area along the Ship Channel between Bayport and Houston, as a result of declining ground-water levels associated with the rapid industrial expansion. During this time, subsidence problems took on crisis proportions, prompting the creation of the Harris-Galveston Coastal Subsidence District. By 1979 up to 10 feet of subsidence had occurred, and almost 3,200 square miles had subsided more than 1 foot.

In the 1940s upstream reservoirs and canals allowed the first deliveries of surface water to Galveston, Pasadena, and Texas City, but ground water remained the primary source until the 1970s. The city of Galveston began converting to surface water supplied from Lake Houston in 1973, and in the late 1970s the cities of Pasadena and Texas City converted to surface water from Lake Livingston, a reservoir on the Trinity River.

Since the late 1970s subsidence has largely been arrested along the Ship Channel and in the Baytown-LaPorte and Pasadena areas due to a reduction in ground-water pumpage made possible by the conversion from ground-water to surface-water supplies. By 1995, total annual ground-water pumpage in the Houston area had declined to only 60 percent of peak amounts pumped during the late 1960s; within the jurisdiction of the Harris-Galveston Coastal Subsidence District, ground-water pumpage constituted only 25 percent of peak amounts. However, as subsidence in the coastal area was stabilizing,

Subsidence trends reflect patterns of resource development that shifted inland from coastal oil and gas extraction to ground-water extraction for municipal and industrial supplies.

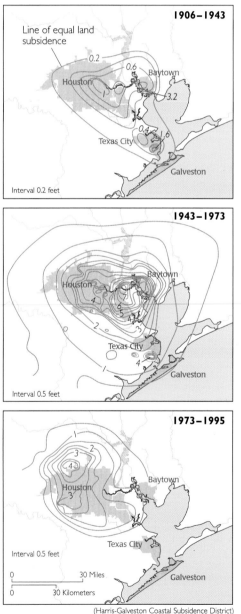

(Harris-Galveston Coastal Subsidence District)

The Harris-Galveston Coastal Subsidence District has arrested subsidence along the western margins of Galveston Bay by substituting imported water for ground water. A new challenge is to manage ground-water use north and west of Houston where water levels are declining and subsidence is increasing.

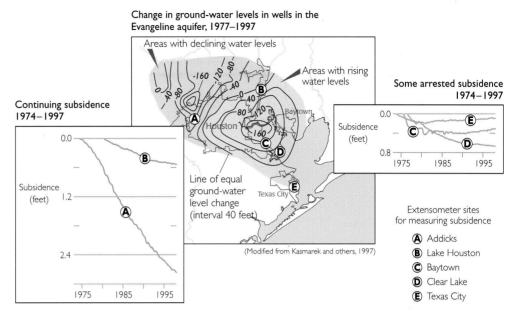

Change in ground-water levels in wells in the Evangeline aquifer, 1977–1997

Areas with declining water levels

Areas with rising water levels

Line of equal ground-water level change (interval 40 feet)

(Modified from Kasmarek and others, 1997)

Continuing subsidence 1974–1997

Some arrested subsidence 1974–1997

Extensometer sites for measuring subsidence

Ⓐ Addicks
Ⓑ Lake Houston
Ⓒ Baytown
Ⓓ Clear Lake
Ⓔ Texas City

subsidence inland—north and west of Houston—was accelerating. In this region ground-water levels have declined more than 100 feet in the Evangeline aquifer between 1977 and 1997, and more than 2.5 feet of subsidence was measured near Addicks between 1973 and 1996.

Texas Gulf Coast Aquifer System

The Evangeline aquifer is the principal source of freshwater

Most of the ground water pumped in the Houston-Galveston area comes from the Chicot and Evangeline aquifers—part of a vast coastal aquifer system that extends throughout the margin of the coastal plain of Texas and Louisiana into Florida. Most of the supply wells are completed in the upper 1,000 to 2,000 feet of the aquifers, where freshwater is available. Saltwater, originally in the aquifers and subsequently flushed by freshwater following sea-level recession, now encroaches on deeper portions of the aquifers. An interface between the saltwater and the overlying freshwater slopes landward from the Galveston coast. Historically, saltwater encroachment in both aquifers has been exacerbated by lowered ground-water levels, especially near the coast. Ground-water quality, levels, and aquifer-system compaction are being closely monitored to minimize any detrimental effects related to overdrafting the ground-water supply.

The Evangeline is recharged directly by precipitation and surface runoff where it crops out north of Houston.

A weak hydraulic connection between shallow ground water, the Chicot aquifer, and the Evangeline aquifer allows the vertical movement of water into and between the aquifers.

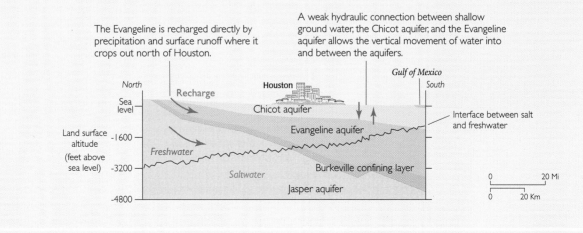

In 1983 Brownwood was flooded after hurricane Alicia produced a storm surge up to 11 feet.

(Harris-Galveston Coastal Subsidence District)

Subsidence increases the frequency and intensity of flooding

Located along a low-lying coast that is subject to tropical storms, the Houston area is naturally vulnerable to flooding. In coastal areas, subsidence has increased the amount of land subject to the threat of tidal inundation. Flooding by tidal surges and heavy rains accompanying hurricanes may block evacuation routes many hours before the storms move inland, endangering inhabitants of islands and other coastal communities. The increased incidence of flooding in coastal areas eventually led to the growing public awareness of subsidence and its costs.

The fate of the Brownwood subdivision of Baytown affords a particularly dramatic example of the dangers of coastal subsidence. Brownwood was constructed, beginning in 1938, as an upper-income subdivision on wooded lots along Galveston Bay (Holzschuh, 1991). At that time the area was generally 10 feet or less above sea level. By 1978 more than 8 feet of subsidence had occurred.

"The subdivision is on a small peninsula bordered by three bays. [It] is a community of about 500 single-unit family houses. Because of subsidence, a perimeter road was elevated in 1974 to allow ingress and egress during periods of normal high tide [about 16 inches], and to provide some protection during unusual high tide. Pumps were installed to remove excess rainfall from inside the leveed area. Because of subsidence after the roadway was elevated, tides of about [4 feet] will cause flow over the road. The United States Army Corps of Engineers studied methods to protect the subdivision from flooding. The cost of a levee system was estimated to be about $70 million. In 1974, the Army Corps estimated that it would cost about $16 million to purchase 442 homes, relocate 1,550 people, and convert [750 acres] of the peninsula into a park. This proposed solution was approved by the Congress of the United States and provided necessary funding. However, the project required that a local sponsor (the City of Baytown) should approve the project, provide 20 per cent of the funds ($3 million) and agree to maintain the park. By the time the first election to fund the project was held on 23 July 1979, the cost estimate had increased to $37.6 million, of which the local share was $7.6 million. The proposal was defeated, and two days later 12 inches of rain fell on Brownwood causing the flooding of 187 homes. Another bond election was held on 9 January 1980 and again the proposal was defeated. Accepting the residents' decision, Baytown officials began planning the sale of $3.5 million worth of bonds to finance the first stage of a fifteen-year, $6.5-million programme to upgrade utilities in the subdivision. Meanwhile, those who own the houses generally also owe mortgages and cannot afford to purchase other homes. Although they continue to live in the subdivision many have to evacuate their homes about three times each year."

—Gabrysch, 1983

Water from Galveston Bay inundated subsiding land and flooded homes in Baytown (1960).

The year that article was published, Hurricane Alicia struck a final blow to Brownwood. All homes in the subdivision were abandoned. Today, most of the subdivision is a swampy area well-suited for waterfowl; egrets and scarlet ibis are often seen.

An abandoned house in the
Brownwood subdivision

Subsidence also exposes inland areas to increased risks of flooding and erosion by altering natural and engineered drainageways (open channels and pipelines) that depend on gravity-driven flow of storm-runoff and sewerage. Differential subsidence, depending on where it occurs with respect to the location of drainageways, may either reduce or enhance preexisting gradients. Gradient reductions decrease the rate of drainage and thereby increase the chance of flooding by storm-water runoff. Gradient reversals may result in ponding or backflow of sewage and stormwater runoff. In some areas, the drainage gradients may be enhanced and the rate of drainage may be increased. In terms of flooding risk, this may have a beneficial effect locally but an adverse effect downstream. For open channels, the changing gradients alter streamflow characteristics leading to potentially damaging consequences of channel erosion and sediment deposition.

Wetlands are being lost to subsidence

Galveston Bay is one of the most significant bay ecosystems in the Nation. The estuary is Texas' leading bay fishery and supports vibrant recreation and tourism industries. Sixty-one percent of the Bay's 232 miles of shoreline is composed of highly productive fringing wetlands but, mainly because of subsidence, more than 26,000 acres of emergent wetlands have been converted to open water and barren flats (White and others, 1993). Subsidence has also contributed to a significant loss of submerged aquatic vegetation (mostly seagrass) since the 1950s. Some bay shorelines have become more susceptible to erosion by wave action due to loss of fringing wetlands. At the same time, the reduction in sediment inflows to the bay system resulting from construction of reservoirs along tributary rivers slows the natural rebuilding of shorelines. Because of the combined and interrelated effects of relative sea-level rise, loss of wetlands, and reduced sediment supply, the shoreline is eroding at an average rate of 2.4 feet per year (Paine and Morton, 1986). As the water level rises, marsh along the shoreline is drowned. When residential, commercial, or industrial development is located near the shoreline, the potential for the landward migration of marshes is eliminated. The result is a reduction in wetland habitats, which provide the foundation for commercial and recreational fisheries.

The most extensive changes in wetlands have occurred along the lower reaches of the San Jacinto River near its confluence with Buffalo Bayou. This area had subsided by 3 feet or more by 1978, resulting in submergence and changes in wetland environments that progressed inland along the axis of the stream valley. Open water displaced riverine woodlands and swamps. Trends along the lower reaches of other rivers, bayous, and creeks have been similar, resulting in an increase in the extent of open water, loss of inland marshes

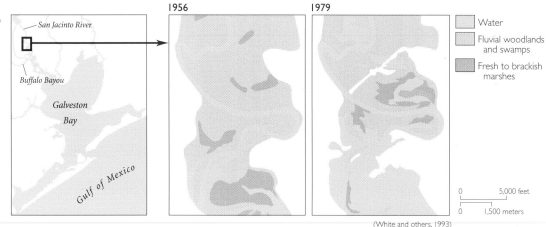

Wetlands were lost to inundation resulting from subsidence in the lower reaches of the San Jacinto River.

1956   1979

Water
Fluvial woodlands and swamps
Fresh to brackish marshes

0   5,000 feet
0   1,500 meters

(White and others, 1993)

and woodlands and, in some areas, the development of new marshes inland from the encroaching waters.

The health and productivity of the bay ecosystem depends on the presence of key habitats like salt marshes, but also on the mix of river and bay water. Many species of fish, wildlife, aquatic plants, and shellfish in Galveston Bay depend on adequate freshwater inflows for survival. The estuary is adapted to highly variable inflows of freshwater. For instance, oysters prefer somewhat salty water, but need occasional surges of freshwater. The volume, timing, and quality of freshwater inflows to the estuary are key factors.

The increasing demand for surface-water supplies, motivated in recent years by efforts to mitigate land subsidence, has led to construction of reservoirs and diversions that have reduced the sediments and nutrients transported to the bay system (Galveston Bay National Estuary Program, 1995). Controlled releases from surface impoundments such as Lake Livingston and Lake Houston have changed the natural freshwater inputs to the bay system; the high flows are lower, the low flows are higher, and peak flows are delayed by about 1 month. As a result, the amount of mineral sediment being delivered by streams to the wetlands has been reduced, limiting some of the natural accretion of wetlands.

Normally, the process of wetland accretion is self-regulated through negative feedback between the elevation of the wetland and relative sea level. When wetland elevations are in balance relative to mean sea level, periodic and frequent tidal inundations mobilize sediment and nutrients in the wetland in a way that favors vegetative growth and a balance between sediment deposition and erosion. Subsidence may upset this balance by submerging the wetland. The drowned wetland cannot support the same floral community, loses its ability to trap sediment as before, and is virtually unregulated by relative sea-level changes. These changes impact the natural processes in the bay and related ecosystems, which evolved with the rhythm of the unregulated streams and rivers.

Coastal subsidence allows shorelines to move landward causing the demise of some coastal woodlands.

(Galveston Bay Information Center, TAMUG)

Subsidence activates faults
Fault creep related to water-level declines

A house in Baytown near Brownwood was damaged by fault creep.

(Holzer and Gabrysch, circa 1987)

Many faults exist in the Houston-Galveston area, both regional-scale "down-to-the-coast" faults that represent slow sliding of the land mass towards the Gulf of Mexico and local structures associated with oil fields (see sidebar on the Goose Creek oil field) (Holzshuh, 1991). Since the late 1930s, 86 active faults with an aggregate scarp length of about 150 miles have offset the land surface and damaged buildings and highways in the metropolitan area (Holzer and Gabrysch, 1987). The scarps typically grow by seismic creep at rates of up to 1 inch per year (Holzer, 1984). Monitoring of fault creep, water levels, and land subsidence has demonstrated a clear cause-and-effect relation. The fault movement is caused by water-level decline and associated subsidence. In the 1970s, a period of water-level recovery began in the eastern part of the Houston area, due to delivery of imported surface water and associated reduction of ground-water pumpage. Fault creep stopped or slowed in the area of water-level recovery, but continued unabated in the area of ongoing water-level decline.

Vertical displacements at eight selected fault-monitoring sites in the Houston area show a pattern related to water-level declines and land subsidence.

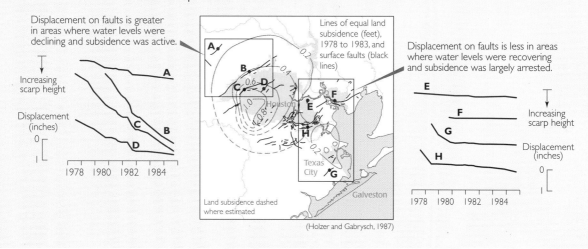

Displacement on faults is greater in areas where water levels were declining and subsidence was active.

Displacement on faults is less in areas where water levels were recovering and subsidence was largely arrested.

(Holzer and Gabrysch, 1987)

SUBSIDENCE IS ACTIVELY MANAGED

Public awareness of subsidence and its causes increased along with the frequency of coastal flooding. In the late 1960s groups of citizens began to work for a reduction in ground-water use. State legislators became educated about the problem, and in May 1975 the Texas Legislature passed a law creating the Harris-Galveston Coastal Subsidence District, the first district of its kind in the United States. The unprecedented Subsidence District was authorized as a regulatory agency, with the power to restrict ground-water withdrawal by annually issuing or denying permits for large-diameter wells, but was forbidden to own property such as water-supply and conveyance facilities.

Increasing ground-water pumpage landward, west and north of Houston, has caused additional, ongoing subsidence.

In areas to the east and south of Houston, regulatory action by the Harris-Galveston Coastal Subsidence District has reduced ground-water pumpage, thus dramatically slowing subsidence.

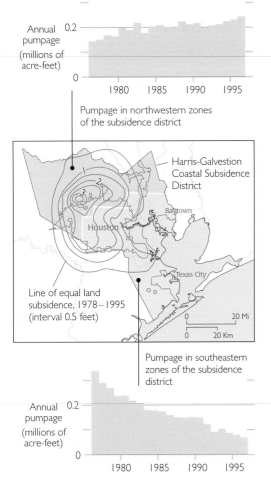

(Harris-Galveston Coastal Subsidence District)

The initial (1976) Subsidence District plan recognized the critical situation in the coastal areas and was designed to have an immediate impact there. Surface water from the recently completed Lake Livingston reservoir on the Trinity River was used to convert industry along the Houston Ship Channel from ground water to surface water. Subsidence in the Baytown-Pasadena area soon slowed dramatically. Earlier imports of surface water from Lake Houston on the San Jacinto River, to the east side of Houston, had locally and temporarily halted water-level declines, but were insufficient to keep pace with the growing demand. The additional water supplied from Lake Livingston was sufficient to significantly reduce ground-water use and ultimately did lead to a recovery of water levels over a large area.

In the eastern part of the greater Houston region, near the bay system, subsidence has been controlled by conversion from ground-water to imported surface-water. However, subsidence is accelerating to the west, where ground-water use has increased. Thus, the area of active subsidence has shifted from the low-lying, tide-affected areas towards higher elevations inland.

A devastating flood in 1984 on Brays Bayou, a major watershed in southwest Houston, renewed concern about the effects of subsidence in inland areas. It was recognized that flood control and subsidence control should be coordinated to minimize flood damages. During the 1989 legislative session, the Fort Bend Subsidence District was created to manage and control subsidence in Fort Bend County.

In 1992, the Harris-Galveston Coastal Subsidence District adopted a regulatory action plan to reduce ground-water pumpage by 80 percent no later than the year 2020. Due to the high cost of constructing distribution lines westward across the metropolitan area, the plan was to be implemented in phases, allowing time to design, finance, and construct surface-water importation facilities. The two subsidence districts will cooperate to ensure coordinated planning of the conversion from ground water to surface water.

The direct and indirect costs of subsidence

The low elevation, proximity to bays and the Gulf of Mexico, dense population, and large capital investment make it likely that the Houston-Galveston area has been more significantly impacted by subsidence than any other metropolitan area in the United States. The actual economic cost of subsidence is hard to quantify, and most published estimates are necessarily vague. For example, Gabrysch (1983) stated that "many millions of dollars" have been spent reclaiming land submerged by tidal water, elevating structures such as buildings, wharves and roadways, and constructing levees to protect against tidal inundation; further, "millions of dollars" are spent on repairing damage due to fault movement. One conservative estimate for the period 1969 to 1974 placed the average annual cost to property owners at more than $31,000,000 in 1975 dollars (Jones, 1976) or about $90,000,000 in 1998 dollars.

After the completion of Lake Houston in 1954, water distribution lines were constructed to convey surface water from Lake Houston to the Pasadena industrial area in order to supplement local ground-water supplies.

(Harris-Galveston Coastal Subsidence District)

The costs of such subsidence-related phenomena as the loss of wetlands are even more difficult to assess than property losses. Although some estimates could be made based on the changing value of commercial and recreational fisheries, it would be difficult to distinguish the influence of subsidence from that of other factors. Similarly, some fraction of the ongoing cost of flood prevention and flood-damage repair could fairly be attributed to subsidence.

The most definitive published subsidence-damage estimates have to do with the costs of relocating dock facilities, constructing hurricane levees, and rectifying drainage problems at refineries along the Houston Ship Channel. For two refineries alone, the estimated total cost was $120,000,000 in 1976 dollars (Holzschuh, 1991), or about $340,000,000 in 1998 dollars. If these estimates are correct, it seems reasonable to suggest that subsidence-related damage to industrial infrastructure alone may run into the billions of dollars.

Ongoing monitoring will help managers plan for the future

Ongoing patterns of subsidence in the Houston area are carefully monitored. Compaction of subsurface material is measured continuously using 13 borehole extensometers (wells equipped with compaction monitors) at 11 sites throughout the region. Piezometers completed to different depths are used to simultaneously monitor water levels. The decreasing subsidence rates observed at sites in the eastern part of the region are a direct result of reducing local ground-water withdrawals through conversion to imported surface-water supplies. In contrast, measurements from the western part of the region reveal continuing subsidence.

A network of 82 bench marks distributed throughout the two subsidence districts was installed in 1987 for determination of elevation changes using the Global Positioning System (GPS). The bench marks were resurveyed using GPS in 1995. The results of the measurements are the basis for the subsidence measured during the 1987 to 1995 period. Continuous Operating Reference Stations (CORS), used to continuously monitor the elevation of three extensometers with GPS, are being maintained by the Harris-Galveston Coastal Subsidence District under the direction of the National Geodetic Survey (NGS). One of the CORS sites is in the NGS Na-

USGS hydrologist measures water levels at an extensometer site, which also serves as a Continuous Operating Reference Station equipped with a GPS antenna and receiver to continuously monitor land subsidence.

1976

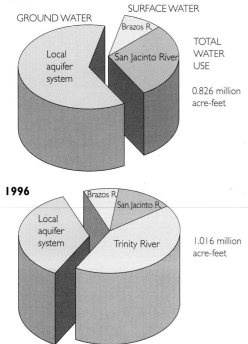

As a percentage of the total, ground-water use has dropped significantly, but total water use is rising.

tional Network. In addition to the fixed locations, portable GPS receivers mounted in trailers are used wherever subsidence measurements are needed. Each portable receiver can operate at up to four different sites each month. GPS is expected to be more cost-effective for monitoring subsidence in the Houston area than constructing additional extensometers or surveying benchmarks using more traditional leveling techniques.

Some controversy attends efforts to gradually achieve conversion to surface water on the north and west sides of Houston, mainly because the imported surface water is expected to cost about twice as much as the ground water that is currently used. Various local municipalities are contesting the timing and apportioning of costs (Houston Chronicle, 27 August 1997, "That sinking feeling hits northwest Houston").

Given the continuing rapid growth of Houston, there is also some long-term concern about securing sufficient surface-water supplies. State and local governments are already at work seeking to ensure that there will be enough water for the expected future population. The primary strategies aim to promote water conservation and acquire supplies from East Texas reservoirs. In addition to the concerns of East Texas communities about water being exported to Houston, such water transfers have ecological effects on the coast and on the waterways through which the water is moved.

The price of water is expected to gradually increase as population and economic growth increase demand. Many farmers will find it difficult to pay higher prices. This may lead to land-use changes in rural communities as farmers find new crops, turn to ranching, or give way to suburban development. Small businesses that support farms will be particularly vulnerable to these changes.

Houston's continuing rapid growth means that subsidence must continue to be vigilantly monitored and managed. However, the region is better-positioned to deal with future problems than many other subsidence-affected areas, for several reasons: a raised public consciousness, the existence of well-established subsidence districts with appropriate regulatory authority, and the knowledge base provided by abundant historical data and ongoing monitoring.

Galveston at sunset

(Harris-Galveston Coastal Subsidence District)

LAS VEGAS, NEVADA

Gambling with water in the desert

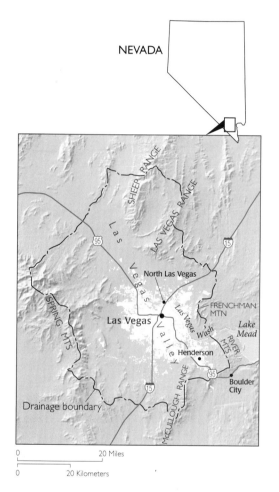

Las Vegas Valley is the fastest growing metropolitan area in the United States (U.S. Department of Commerce, accessed July 27, 1999). The accelerating demand for water to support the rapid growth of the municipal-industrial sector in this desert region is being met with imported Colorado River System supplies and local ground water. The depletion of once-plentiful ground-water supplies is contributing to land subsidence and ground failures. Since 1935, compaction of the aquifer system has caused nearly 6 feet of subsidence and led to the formation of numerous earth fissures and the reactivation of several surface faults, creating hazards and potentially harmful impacts to the environment.

In the near future, the current water supplies are expected not to satisfy the anticipated water demand. The federally mandated limit placed on imported water supplied from nearby Lake Mead, a reservoir on the Colorado River, will likely force a continued reliance on ground water to supplement the limited imported-water supplies. Water supply-and-demand dynamics in this growing desert community will likely perpetuate problems of land subsidence and related ground failures in Las Vegas Valley, unless some balanced use of the ground-water resource can be achieved.

Michael T. Pavelko, David B. Wood, and Randell J. Laczniak
U.S. Geological Survey, Las Vegas, Nevada

"THE MEADOWS" WAS AN IMPORTANT DESERT OASIS

Wednesday Oct. 11th 1848

[…] Camped about midnight at a spring branch called Cayataus. Fair grass. This is what is called the "Vegas".

Thursday Oct. 12th 1848

[…] Staid [sic] in the camp we made last night all day to recruit the animals. They done finely. There is the finest stream of water here, for its size, I ever saw. The valley is extensive and I doubt not [,] would by the aid of irrigation be highly productive. There is water enough in this rapid little stream to propel a grist mill with a dragger run of stones! And oh! such water. It comes, too, like an oasis in the desert, just at the termination of a 50 m. [mile] stretch without a drop of water or a spear of grass. […]"

Orville C. Pratt (from The Journal of
Orville C. Pratt, 1848 in Hafen and
Hafen, 1954)

Las Vegas Valley is located in southern Nevada and lies within both the Great Basin and Mojave Desert sections of the Basin and Range physiographic province. The arid, northwest-trending valley is bounded on the west by several mountain ranges and drains a 1,564-square-mile watershed southeastward through Las Vegas Wash into Lake Mead.

More than 24 inches of precipitation fall annually in the Spring Mountains bounding the valley to the west, but less than 4 inches of rain fall annually on the valley floor; measurable amounts (greater than 0.01 inch) seldom occur more than 30 days each year. Temperatures range from below freezing in the mountains to more than 120° F on the valley floor. There are typically more than 125 days of 90° F or warmer temperatures each year in Las Vegas Valley (Houghton and others, 1975).

The desert oasis of Las Vegas Valley has been a source of water for humans for more than 13,000 years. Native Americans of the Mojave and Paiute tribes were among the earliest known users. Named by an unknown trader for its grassy meadows, Las Vegas, Spanish for "the meadows," was a watering stop along the Old Spanish Trail that connected the settlements in Los Angeles and Santa Fe. In 1844, the renowned explorer John C. Fremont stopped here and spoke of the waters as "two narrow streams of clear water, 4 or 5 feet deep, with a quick current, from two singularly large springs" (Mendenhall, 1909). Others were similarly moved by the refreshing contrast of these welcome meadows in the otherwise barren landscape.

The railroad initiates a period of rapid growth

After failed attempts by Mormon settlers to mine lead from the nearby Spring Mountains and to establish farming in the valley, a flourishing ranch supported by springs and Las Vegas Creek was established in 1865 by Octavius Decatur Gass, a settler who had initially been attracted to the West by gold mining. In 1905, Montana Senator William Clark brought the San Pedro, Los Angeles and Salt Lake Railroad to the valley and established the small town of Las Vegas, a site chosen because of its central location between Los Angeles and Salt Lake City, and because of the water supply necessary to keep the steam locomotives running.

Fremont Street, Las Vegas, looking west (ca. 1910)

(Junior League of Las Vegas Collection, University of Nevada, Las Vegas Library)

The Las Vegas Land and Water Company, established in 1905, was the area's first water purveyor.

As the railroad grew, so did Las Vegas and its thirst for water (Jones and Cahlan, 1975). To help meet the increasing demand, the Las Vegas Land and Water Company was formed in 1905. A new period of growth began in 1932 with the construction of Boulder Dam (later renamed Hoover Dam) and Lake Mead on the Colorado River, southeast of Las Vegas. Boulder Dam brought workers to Las Vegas from throughout America, and provided a seemingly unlimited supply of water and power in one of the most unlikely places. The wealth of land, water, and power resources attracted industry, the military, and gambling to the valley during the 1940s and 1950s. The population of Las Vegas was growing steadily, and by 1971 the heightened water demand required importing additional water from Lake Mead through a newly constructed Southern Nevada Water Project pipeline. At present, Las Vegas Valley is home to 1.2 million people, about two-thirds of Nevada's population, and hosts more than 30 million tourists each year.

Today Las Vegas sprawls across the valley.

Urban growth in the Las Vegas Valley has soared in the last few decades.

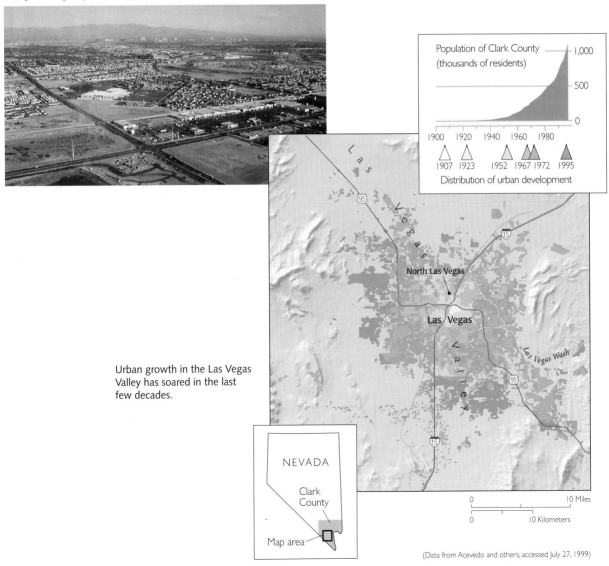

(Data from Acevedo and others, accessed July 27, 1999)

In 1912, the Eglington well, one of several uncapped artesian wells, was allowed to flow freely. (It is shown here flowing at about 615 gallons per minute.)

(Carpenter, 1915)

By 1938 the Eglington well had ceased flowing. The water level was then 3.3 feet below land surface.

(Livingston, 1941)

BROWNING OF "THE MEADOWS": DEMAND FOR WATER DEPLETES THE AQUIFER SYSTEM

Prior to development in Las Vegas Valley, there was a natural, albeit dynamic, balance between aquifer-system recharge and discharge. Over the short term, yearly and decadal climatic variations (for example, drought and the effects of El Niño) caused large variations in the amount of water available to replenish the aquifer system. But over the long term, the average amount of water recharging the aquifer system was in balance with the amount discharging, chiefly from springs and by evapotranspiration. Estimates of the average, annual, natural recharge of the aquifer system range from 25,000 to 35,000 acre-feet (Maxey and Jameson, 1948; Malmberg, 1965; Harrill, 1976; Dettinger, 1989).

In 1907, the first flowing well was drilled by settlers to support the settlement of Las Vegas, and there began to be more ground-water discharge than recharge (Domenico and others, 1964). Uncapped artesian wells were at first permitted to flow freely onto the desert floor, wasting large quantities of water. This haphazard use of ground water prompted the State Engineer, W.M. Kearney, to warn in 1911 that water should be used "… with economy instead of the lavish wasteful manner, which has prevailed in the past" (Maxey and Jameson, 1948).

Intensive ground-water use led to steady declines in spring flows and ground-water levels throughout Las Vegas Valley. Spring flows began to wane as early as 1908 (Maxey and Jameson, 1948). By 1912 nearly 125 wells in Las Vegas Valley (60 percent of which were flowing-artesian wells) were discharging nearly 15,000 acre-feet per year.

Las Vegas' water supply has kept pace with the demand.

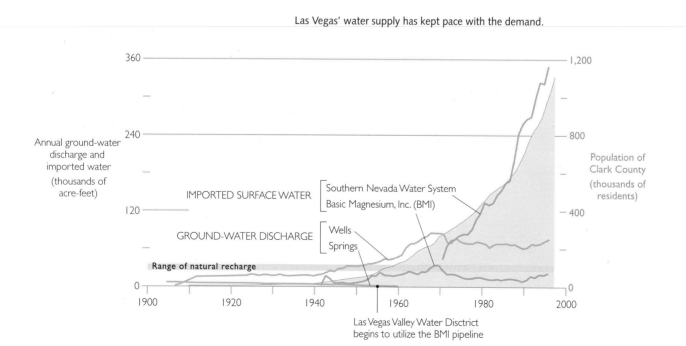

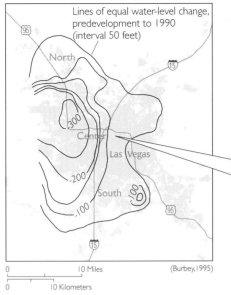

Lines of equal water-level change, predevelopment to 1990 (interval 50 feet)

(Burbey, 1995)

By 1990 areas of the valley that had once supported flowing artesian wells experienced water level declines of more than 300 feet.

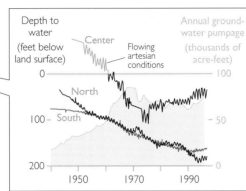

Increasing pumpage through the 1960s caused water levels to drop throughout Las Vegas Valley. Presently, due to some stabilization in the pumpage amounts and artificial ground-water recharge programs, water levels are recovering in many areas of the valley.

With the construction of Boulder Dam came development of the military and industrial sectors and a rapidly increasing demand for water. In 1942 a water pipeline was constructed to bring water from Lake Mead to the Basic Magnesium Project (now called Basic Management, Inc.) in the City of Henderson. This pipeline marked the first supplementation of Las Vegas Valley ground water and the beginning of surface-water imports to the valley. In 1955 the Las Vegas Valley Water District (LVVWD) began to use this pipeline to supplement the growing water demands. By this time, the amount of ground water pumped annually from wells had reached nearly 40,000 acre-feet, surpassing the estimated natural recharge to the valley aquifer system (Mindling, 1971). By 1968 the annual ground-water pumpage in the valley reached nearly 88,000 acre-feet (Harrill, 1976).

In 1971, the capacity to import surface water into the valley was greatly expanded when a second, larger pipeline was constructed between Lake Mead and Las Vegas by the Southern Nevada Water Project (Harrill, 1976). However, despite the steady increases in imported surface-water deliveries, rising demand for water and federally stipulated limits on Lake Mead imports encouraged a continued dependence on the local ground-water resource.

Ground-water levels decline as Las Vegas expands

Between 1912 and 1944, ground-water levels declined at an average rate of about 1 foot per year (Domenico and others, 1964). Between 1944 and 1963, some areas of the valley experienced water-level declines of more than 90 feet (Bell, 1981a). The City of North Las Vegas was the first area to experience large water-level declines but, as Las Vegas expanded, new wells were drilled, pumping patterns changed, and ground-water-level declines spread to areas south and west of the City of North Las Vegas. Between 1946 and 1960, the area of the

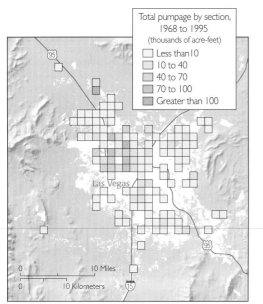

Total pumpage by section,
1968 to 1995
(thousands of acre-feet)

Less than 10
10 to 40
40 to 70
70 to 100
Greater than 100

0   10 Miles
0   10 Kilometers

(Data compiled from unpublished Las Vegas Valley water
usage reports, Nevada Department of Conservation
and Natural Resources, Divison of Water Resources)

valley that could sustain flowing-artesian wells shrank from more than 80 square miles (Maxey and Jameson, 1948) to less than 25 square miles (Domenico and others, 1964). By 1962, the springs that had supported the Native Americans, and those who followed, were completely dry (Bell, 1981a).

Since the 1970s annual ground-water pumpage in the valley has remained between 60,000 and 90,000 acre-feet; most of that has been pumped from the northwestern part of the valley. By 1990 areas in the northwest experienced more than 300 feet of decline, and areas in the central (including downtown and The Strip) and southeastern (Henderson) sections experienced declines between 100 and 200 feet (Burbey, 1995).

In 1996, imports from Lake Mead provided Las Vegas Valley with approximately 356,000 acre-feet of water (Coache, 1996) and represented the valley's principal source of water. This amount included 56,000 acre-feet of return-flow credits for annual streamflow discharging into Lake Mead from Las Vegas Wash.

DEPLETION OF THE AQUIFER SYSTEM CAUSES SUBSIDENCE

Land subsidence and related ground failures in Las Vegas Valley were first recognized by Maxey and Jameson (1948) based on comparisons of repeat leveling surveys made by the USGS and the U.S. Coast and Geodetic Survey between 1915 and 1941. Since then, repeat surveys of various regional networks have shown continuous land subsidence throughout large regions within the valley.

The surveys have revealed that subsidence continued at a steady rate into the mid-1960s, after which rates began increasing through 1987 (Bell, 1981a; Bell and Price, 1991). Surveys made in the 1980s delineate three distinct, localized subsidence bowls, or zones, superimposed on a larger, valley-wide subsidence bowl. One of these smaller subsidence bowls, located in the northwestern part of the valley, subsided more than 5 feet between 1963 and 1987. Two

1964

1997

These photographs of a protruding well just west of downtown Las Vegas show evidence of subsidence. The 1964 photograph shows that the ground has subsided enough, relative to the well casing, to suspend the broken concrete foundation of the well head above land surface. Thirty three years later well head protrudes farther as the ground has continued to subside.

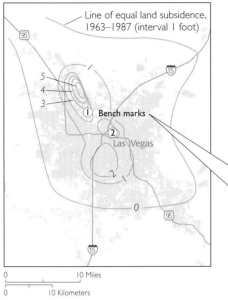

Line of equal land subsidence, 1963–1987 (interval 1 foot)

Bench marks

Las Vegas

0   10 Miles

0   10 Kilometers

Three subsidence bowls were identified between 1963 and 1987. These bowls are caused by a combination of ground-water declines and the presence of compressible sediments in the aquifer system at these locations.

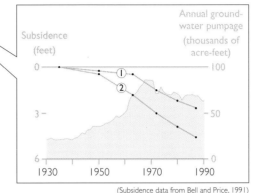

Subsidence measured at two bench marks continued beyond 1970, although ground-water pumpage was slightly reduced.

(Subsidence data from Bell and Price, 1991)

other localized subsidence bowls, in the central (downtown) and southern (Las Vegas Strip) parts of the valley, subsided more than 2.5 feet between 1963 and 1987. The areas of maximum subsidence do not necessarily coincide with areas of maximum water-level declines. One likely explanation is that those areas with maximum subsidence are underlain by a larger aggregate thickness of fine-grained, compressible sediments (Bell and Price, 1991).

Aquifer-system compaction creates earth fissures and reduces storage

All the impacts of subsidence in Las Vegas Valley have not yet been fully realized. Two important impacts that have been documented are (1) ground failures—localized ruptures of the land surface; and (2) the permanent reduction of the storage capacity of the aquifer system. Other potential impacts that have not been studied extensively are:

• Creation of flood-prone areas by altering natural and engineered drainage ways;

• Creation of earth fissures connecting nonpotable or contaminated surface and near-surface water to the principal aquifers; and

• Replacement costs associated with protruding wells and collapsed well casings and well screens.

All of these potential damages create legal issues related to mitigation, restoration, compensation, and accountability.

Ground failures Earth fissures are the dominant and most spectacular type of ground failure associated with ground-water withdrawal in Las Vegas Valley. Earth fissures are tensile failures in subsurface materials that result when differential compaction of sediments pulls apart the earth materials. Buried, incipient earth fissures be-

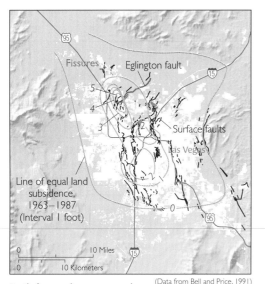

Line of equal land subsidence, 1963–1987 (Interval 1 foot)

(Data from Bell and Price, 1991)

Earth fissures have occurred near areas of greater differential subsidence, and many fissures are associated with surface faults.

come obvious only when they breach the surface and begin to erode, often following extreme rains or surface flooding conditions. Earth fissures have been observed in Las Vegas Valley as early as 1925 (Bell and Price, 1991), but were not linked directly to subsidence until the late 1950s (Bell, 1981a). Most of the earth fissures are areally and temporally correlated with ground-water level declines.

Movement of preexisting surface faults has also been correlated to ground-water level changes and differential land subsidence in numerous alluvial basins (Holzer, 1979; Bell, 1981a; Holzer, 1984). In Las Vegas Valley, earth fissures often occur preferentially along preexisting surface faults in the unconsolidated alluvium. They tend to form as a result of the warping of the land surface that occurs when the land subsides more on one side of the surface fault than the other. This differential land subsidence creates tensional stresses that ultimately result in fissuring near zones of maximum warping. The association of most earth fissures with surface faults suggests a causal relationship. The surface faults may act as partial barriers to ground-water flow, creating a contrast in ground-water levels across the fault, or may offset sediments of differing compressibility.

The associated land-surface displacements and tilts are often sufficient to damage rigid or precisely leveled structures. Damage to homes in a 241-home subdivision in the north-central part of the valley has already cost more than $6 million, and the total cost projections are in excess of $14 million (Marta G. Brown, City of North Las Vegas, written communication, 1997). Other damage related to fissuring includes cracking and displacement of roads, curbs, sidewalks, playgrounds, and swimming pools; warped sewage lines; ruptured water and gas lines; well failures resulting from shifted, sheared, and/or protruded well casings; differential settlement of railroad tracks; and a buckled drainage canal (Bell, 1981b; Marta G. Brown, City of North Las Vegas, written communication, 1997). Earth fissures are also susceptible to erosion and can form wide, steep-walled gullies capable of redirecting surface drainage and creating floods and other hazards. Adverse impacts of ground failures may worsen as the valley continues to urbanize and more developed areas become affected.

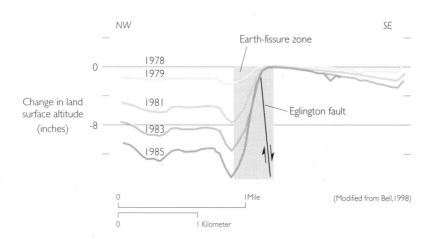

This cross section of the Eglington fault zone and accompanying fissure zone shows that land-surface elevations on the upthrown side of the fault are decreasing due to subsidence.

A fissure displaces pavement (far right) and damages a building (near right) on Harrison Street, Las Vegas.

(Fred B. Houghton, 1961)

An estimated 187,000 acre-feet (61 billion gallons) of water (enough water to supply almost 10,000 households in Las Vegas for nearly 20 years) may have been derived from a permanent reduction in the storage capacity of the Las Vegas Valley aquifer system due to compaction of the aquifer system and land subsidence between 1907 and 1996.

Reduced storage capacity Reduction of storage capacity in the Las Vegas Valley aquifer system is another important consequence of aquifer-system compaction. The volume of ground water derived from the irreversible compaction of the aquifer system —"water of compaction"—is approximately equal to the reduced storage capacity of the aquifer system and represents a one-time quantity of water "mined" from the aquifer system.

Loss of aquifer-system storage capacity is cause for concern, especially for a fast-growing desert metropolis that must rely in part on local ground-water resources. A study conducted by the Desert Research Institute (Mindling, 1971) estimated that, at times, up to 10 percent of the ground water pumped from the Las Vegas Valley aquifer system has been derived from water of compaction. Assuming conservatively that only 5 percent of the total ground water pumped between 1907 and 1996 was derived from water of compaction, the storage capacity of the aquifer system has been reduced by about 187,000 acre-feet. This may or may not be considered "lost" storage capacity: arguably, if this water is derived from an irreversible process, this storage capacity has been used in the only way that it could have been. In any case, producing water of compaction represents mining ground water from the aquifer system. Further, the reduced storage implies that, even if water levels recover completely, any future drawdowns will progress more rapidly.

LAS VEGAS VALLEY IS UNDERLAIN BY A GROUND-WATER RESOURCE

Las Vegas Valley is a sediment-filled structural trough that has formed over many millions of years through compression, extension, and faulting of the original flat-lying marine sediments that form the bedrock. Some bedrock blocks were down-dropped between the faults along the eastern and western margins of the present-day valley.

Sediment eroded by wind and water from the surrounding bedrock highlands began filling the trough with gravel, sand, silt, and clay.

During some of the wetter periods in the past 1 million years or so, extensive playa lakes and spring-fed marshes covered the lower parts of the valley floor, depositing variably thick sequences of fine-grained sediment (Mifflin and Wheat, 1979 and Quade et al., 1995). Coarse-grained sand and gravel tend to rim the valley, forming alluvial fans and terraces, especially in the northern, western, and southern parts. The deposits generally thicken and become finer-textured toward the central and eastern part of the valley, where their total thickness exceeds 5,000 feet (Plume, 1989).

Ground water flows through the aquifers

Ground water is generally pumped from the upper 2,000 feet of unconsolidated sediments that constitute the aquifer system in the central part of the valley. The deeper aquifers, generally below 300 feet, are capable of transmitting significant quantities of ground water, and have been referred to variously as the "principal," "artesian," or "developed-zone" aquifers (Maxey and Jameson, 1948; Malmberg, 1965; Harrill,1976; Morgan and Dettinger, 1996). In places, these principal aquifers are more than 1,000 feet thick and consist mainly of sands and gravels beneath the terraces along the margins of the valley. In the central and eastern parts, clays and silts predominate (Plume, 1989). Overlying the principal aquifers, in most places, is a 100-to-300 foot-thick section of extensive clay, sand, and gravel deposits known as the "near-surface reservoir." The principal aquifers and the near-surface reservoir are separated by a variably-thick, laterally discontinuous aquitard, or confining unit.

Much of the ground water found in the aquifer system originates as rain or snow falling on the Spring Mountains to the west or on the Sheep and Las Vegas Ranges to the northwest. Some of the precipitation infiltrates into the underlying bedrock through faults and fractures, eventually moving into the deposits comprising the principal aquifers. The remainder of the precipitation runs off onto the sloping alluvial terraces and rapidly enters the sand and gravel deposits, where it either recharges the underlying principal aquifers or is evaporated or transpired into the atmosphere.

Near the margins of the valley, ground water moves freely through the coarse-grained sand and gravel deposits, but as it moves

Most precipitation in the watershed falls in the mountains surrounding Las Vegas

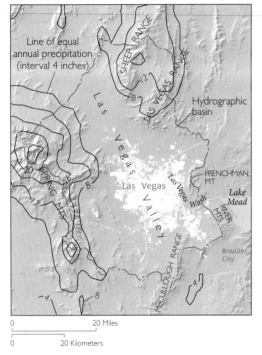

"The settlement [subsidence] in Las Vegas Valley as a whole appears to be the result of compaction of the sediments of the valley fill, and the faults, ... are probably caused by the differential compaction of the fine-grained and coarse-grained sediments."

—1948, George B. Maxey and C. Harry Jameson

Predevelopment

Ground water was sustained by natural recharge, and excess ground water discharged through several springs and into the Las Vegas Wash.

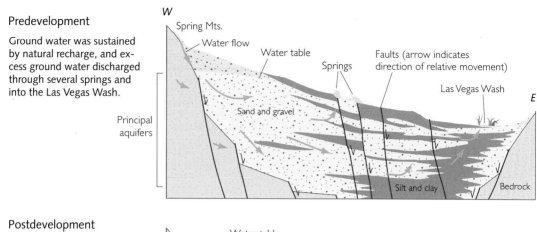

Postdevelopment

Excessive pumping has caused the water table to drop and springs to dry up. Urban run-off has created a reservoir of poorer quality, potentially contaminated water just below the surface that now recharges the principal aquifers.

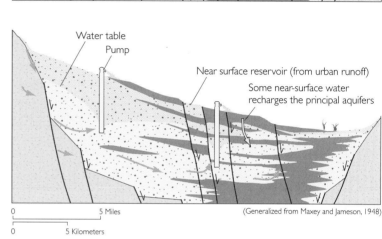

(Generalized from Maxey and Jameson, 1948)

0 ————————— 5 Miles

0 ————————— 5 Kilometers

Vertical exaggeration 15x

basinward it begins to encounter increasingly greater percentages of lower permeability, fine-grained clay and silt. The increasing proportion of fine-grained deposits retards lateral flow, and the low-permeability deposits effectively impede the vertical flow of ground water. As ground water recharges the aquifer system from the higher elevations, fluid pressures in the principal aquifers can build to create artesian conditions at lower elevations in the basin.

Prior to development of the ground-water resource, artesian pressure in the aquifer system forced water slowly upward through confining zones and more rapidly along faults. Flow from these conduits formed the springs on the valley floor and supported thriving grassy meadows with an estimated annual flow of 7,500 acre-feet (Malmberg, 1965). Most of the spring flow and precipitation falling on the valley floor was consumed by evapotranspiration, but some infiltrated downward into the surficial deposits.

The changing balance between recharge and discharge

Development of the ground-water resource to support the local population and its land uses drastically altered the way water cycles through the basin. The present water budget reveals that only a small fraction of the water used in Las Vegas Valley is actually consumed, and therefore removed from the water cycle, by domestic,

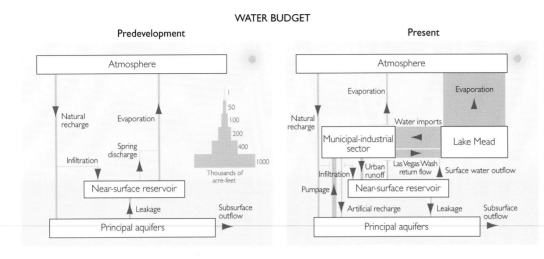

WATER BUDGET

Predevelopment   Present

agricultural and municipal/industrial uses. Most is either returned to the aquifer system, evaporated, or discharged into the Colorado River system. Large quantities of this generally poorer-quality water drain from overwatered lawns, public sewers, paved surfaces, and other drainage ways. Much of this urban runoff flows onto open ground where it evaporates, is transpired by plants, or recharges the near-surface reservoir. Large amounts of treated sewage water are discharged into the Colorado River system by way of the Las Vegas Wash. Ground water has been depleted in the principal aquifers and aquitards, causing land subsidence, while the shallow, near-surface reservoir has been recharged with poor-quality urban runoff.

LAS VEGAS IS DEALING WITH A LIMITED WATER SUPPLY

Managing land subsidence in Las Vegas Valley is linked directly to the effective use of ground-water resources. At present more ground water is appropriated by law and is being pumped in Las Vegas Valley than is available to be safely withdrawn from the ground-water basin (Nevada Department of Conservation and Natural Resources, 1992; Coache, 1996). Historic and recent rates of aquifer-system depletion caused by overuse of the ground-water supply cannot be sustained without contributing further to land subsidence, earth fissures, and the reactivation of surface faults.

"All data available from this and other studies strongly indicate that the quantities of water presently developed, if removed entirely from the ground-water reservoir on a permanent basis, would eventually result in critical depletion"

—*Domenico and others, 1964*

In order to arrest subsidence in the valley, ground-water levels must be stabilized or maintained above historic low levels. Stabilization or recovery of ground-water levels throughout the valley will require that the amount of ground water pumped from the aquifers be less than or equal to the amount of water recharging the system. Eliminating any further decline will reduce the stresses contributing to the compaction of the aquifer system. Even so, a significant amount of land subsidence (residual compaction) will continue to occur until the aquifer system equilibrates fully with the stresses imposed by lowered ground-water levels in the aquifers (Riley, 1969). This equilibrium may require years, decades, or even centuries to be realized.

Replenishing the aquifer system artificially

Las Vegas Valley Water District (LVVWD) and the City of North Las Vegas have developed artificial recharge programs

The artificial recharge programs serve two primary purposes:

- To store surplus imported surface water in the principal aquifers during winter months when demand is relatively low, so that it can later be pumped to supplement any shortfalls in the supply and delivery of imported water during the high-demand summer months

- To replenish the principal aquifers, if only temporarily, thus raising ground-water levels and forestalling subsidence in the local area.

Recharging began in 1988 and by 1995 a total of nearly 115,000 acre-feet of treated, imported Lake Mead water had been injected through more than 40 wells, at an annual rate of up to 25,000 acre-feet. Additional recharge wells constructed since 1995 have significantly enlarged the recharge area and increased the number of injection-well sites.

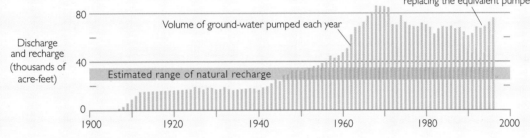

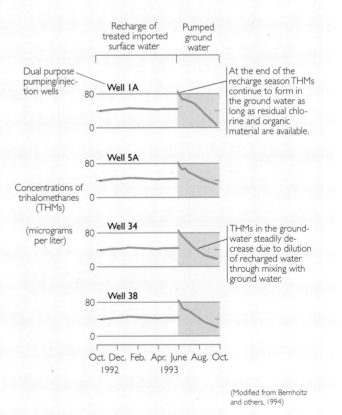

DISINFECTION BYPRODUCTS

The artificial recharge program poses a potential for contamination of the Las Vegas Valley aquifer system. The problem arises because it is necessary to disinfect the recharge water prior to injecting it through the wells into the aquifer system. Disinfection byproducts (DBPs), chiefly trihalomethanes (THMs), form when chlorine is introduced into the water-treatment process. The dissolved and particulate organic material in the water reacts with the chlorine and other halogens to form DBPs, of which THMs are specifically regulated by State and Federal standards. THMs have been shown to cause cancer in laboratory animals, and may pose other health risks to humans. Presently, the total THM maximum contaminant level allowed under the drinking-water standards is 100 µg/l (micrograms per liter), but the U. S. Environmental Protection Agency is strongly considering a lower limit.

Native ground waters in arid alluvial basins are typically low in dissolved organics compared to surface waters, so that even if they are chlorinated prior to use, few if any THMs form. In contrast, the imported surface water is high in organics, and when it is disinfected before injection into the aquifer system, an average of 45 µg/l of THMs are produced. This concentration eventually becomes diluted within the aquifer. But when the mixture is pumped for use, disinfection is still needed, and the chlorine raises THM levels about 25 µg/l, potentially near the drinking-water standard. To lower the THMs to acceptable levels, further treatment or blending (dilution) may be needed.

(Modified from Bernholtz and others, 1994)

The natural recharge is augmented "artificially"

Since 1988, the LVVWD and the City of North Las Vegas have implemented artificial ground-water-recharge programs in an attempt to increase local water supplies during periods of high demand. These aquifer-recharge programs replenish the aquifers by injecting treated surface water imported from Lake Mead through dual-purpose wells. Water is recharged primarily during cooler months, when water demand is lowest, thereby raising ground-water levels above typical winter conditions. Recently, annual artificial recharge of nearly 20,000 acre-feet has succeeded in raising ground-water levels in some local areas to the extent that they are generally higher both at the beginning and end of the peak water-demand (summer) season.

Despite the ambitious efforts to artificially recharge the aquifer system, valleywide net ground-water pumpage still exceeds the estimated natural recharge. To minimize any future subsidence, some combination of increased recharge and reduced pumpage is needed, especially in areas prone to subsidence. These options depend largely on the seasonal availability of additional imported water, to compensate for any additional water recharged, and on the amount of reduced pumpage required to maintain ground-water levels above critical levels.

Both the ground water and surface water of Nevada belong to the public and are managed on their behalf by the State of Nevada, the Colorado River Compact, and the Bureau of Reclamation. Nevada water law is founded on the doctrine of prior appropriation—"first in time, first in right"—which grants the first user of a water course a priority right to the water. All the surface- and ground-water resources in the valley are currently fully appropriated. The State Engineer has established a perennial yield of 25,000 acre-feet for the Las Vegas Valley aquifer system (Malmberg, 1965; Nevada Dept. Of Conservation and Natural Resources, 1992), based on the minimum, average annual natural recharge to the aquifer system. Despite this legally established yield, more than 25,000 acre-feet have been pumped from the valley every year since 1945; a maximum yield of more than 86,000 acre-feet were pumped in 1968. As of

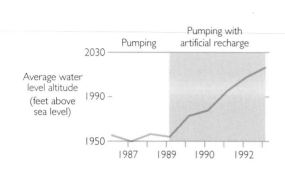

Water levels at the Las Vegas Valley Water District's main well field have increased with artificial recharge.

This typical artificial recharge well has the dual function of pumping and injecting. (The tall object on the far right is the electric motor for the pump).

Water levels and compaction fluctuate seasonally in response to natural and artificial recharge and pumpage.

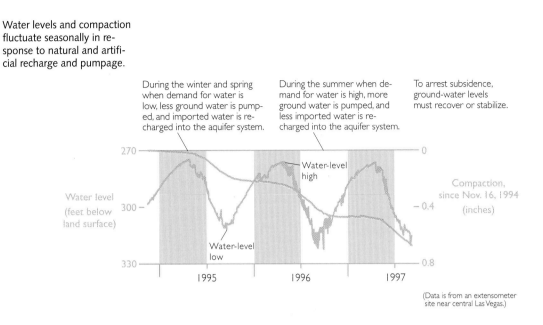

During the winter and spring when demand for water is low, less ground water is pumped, and imported water is recharged into the aquifer system.

During the summer when demand for water is high, more ground water is pumped, and less imported water is recharged into the aquifer system.

To arrest subsidence, ground-water levels must recover or stabilize.

Water level (feet below land surface)

Water-level high

Water-level low

Compaction, since Nov. 16, 1994 (inches)

(Data is from an extensometer site near central Las Vegas.)

1996, State permits for an annual total of 90,000 acre-feet had been issued (Coache, 1996), and in that year nearly 76,000 acre-feet, more than three times the perennial yield, were pumped.

WATER MANAGERS ATTEMPT TO MEET GROWING WATER DEMAND

A limit on the amount of water that can be imported from the Colorado River system, and a growing local water demand, make it difficult to reduce the present reliance on the local ground-water supply. At the current rate of ground-water extraction, there may be insufficient surplus of imported water to control land subsidence. Water-use projections for southern Nevada have indicated that the region's available water supply likely will not meet projected demands beyond the year 2002, or 2006 provided responsible water-conservation programs are implemented (Water Resources Management Incorporated, 1991). After that time, the water supply will become extremely vulnerable to variability caused by droughts and potentially by contamination.

It is uncertain whether Nevada will be able to acquire, on a permanent basis, any additional Colorado River system water beyond the current annual allocation of 300,000 acre-feet. To help prevent water shortages, and thereby reduce additional stress on the aquifer system, the Southern Nevada Water Authority (SNWA) is pursuing several avenues to increase the future supply of water to southern Nevada and Las Vegas Valley. Primary sources might include importation of both in-state and out-of-state water and ground-water banking. Water from the Virgin and Muddy Rivers and ground-water banking in southern Nevada and Arizona are leading options. Stormwater recovery and desalination are also being considered.

Perhaps the most desirable option to the SNWA would be the "wheeling" of Virgin and Muddy River water. Under this scenario, river water that is legally available for use is allowed to continue to flow into Lake Mead, rather than being piped directly out of the rivers. This would allow the SNWA to obtain approximately an additional 120,000 acre-feet, without constructing a pipeline. "Wheeling" of this water, however, is technically not permitted, because any river water that reaches Lake Mead is legally considered to be part of Nevada's Colorado River system water apportionment of 300,000 acre-feet. If legal solutions cannot be achieved in favor of "wheeling" water, a legal, and costly, pipeline could divert this water before it reaches Lake Mead.

Another important potential resource is ground-water banking, whereby aquifers could be artificially recharged with unused portions of Colorado River system water to be used during future high-demand periods. While this option is already being used in Las Vegas Valley, more water could be banked elsewhere in southern Nevada and, pending legal decisions, Nevada could buy water for banking from Arizona or other member states in the Colorado River Compact.

Given these expanded options, the SNWA has projected that there will in fact be enough water to meet the demands of southern Nevada beyond the year 2025.

South-Central Arizona

Earth fissures and subsidence complicate development of desert water resources

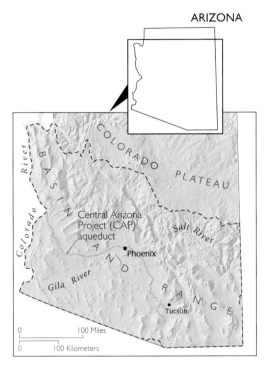

E arth fissures that rupture the Earth's surface and widespread land subsidence in deep alluvial basins of southern Arizona are related to ground-water overdrafts. Since 1900 ground water has been pumped for irrigation, mining, and municipal use, and in some areas more than 500 times the amount of water that naturally replenishes the aquifer systems has been withdrawn (Schumann and Cripe, 1986). The resulting ground-water-level declines—more than 600 feet in some places—have led to increased pumping costs, degraded the quality of ground water in many locations, and led to the extensive and uneven permanent compaction of compressible fine-grained silt- and clay-rich aquitards. A total area of more than 3,000 square miles has been affected by subsidence, including the expanding metropolitan areas of Phoenix and Tucson and some important agricultural regions nearby.

Earth fissures, a result of ground failure in areas of uneven or differential compaction, have damaged buildings, roads and highways, railroads, flood-control structures, and sewer lines. The presence and ongoing threat of subsidence and fissures forced a change in the planned route of the massive, federally-financed Central Arizona Project (CAP) aqueduct that has delivered imported surface water from the Colorado River to central

Michael C. Carpenter
U.S. Geological Survey, Tucson, Arizona

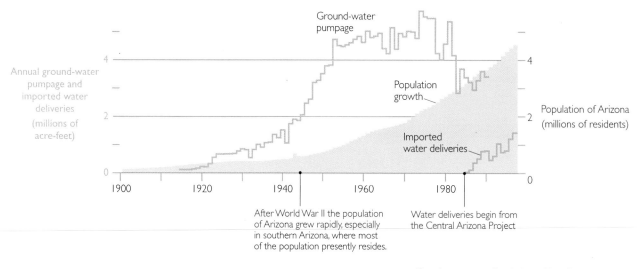

After World War II the population
of Arizona grew rapidly, especially
in southern Arizona, where most
of the population presently resides.

Water deliveries begin from
the Central Arizona Project

(Ground-water pumpage from Anning and Tuet, 1994; imported water
deliveries from Arizona Department of Water Resources; population
data modified from U.S. Census Bureau)

Arizona since 1985. In the CAP, Arizona now has a supplemental
water supply that has lessened the demand and overdraft of
ground-water supplies. Some CAP deliveries have been used in
pilot projects to artificially recharge depleted aquifer systems. When
fully implemented, recharge of this imported water will help to
maintain water levels and forestall further subsidence and fissure
hazards in some areas.

GROUND WATER HAS SUSTAINED AGRICULTURE

Irrigation is needed to grow crops in Arizona because of the low
annual rainfall and the high rate of potential evapotranspiration—
more than 60 inches per year. Precipitation in south-central Arizona
ranges from as low as 3 inches per year over some of the broad flat
alluvial basins to more than 20 inches per year in the rugged mountain ranges. Large volumes of water can be stored in the intermon-
tane basins, which contain up to 12,000 feet or more of sediments
eroded from the various metamorphic, plutonic, volcanic, and consolidated sedimentary rocks that form the adjacent mountains.
Ground water is generally produced from the upper 1,000 to 2,000
feet of the basin deposits, which constitute the aquifer systems.
Ground water pumped from the aquifer systems became a reliable
and heavily tapped source of irrigation water that fueled the development of agriculture during the early and mid-20th century. In
many areas, the aquifer systems include a large fraction of fine-
grained deposits containing silt and clay that are susceptible to compaction when the supporting fluid pressures are reduced by pumping.

CAP water sustains urban growth

Pumping for irrigation began prior to 1900, and increased markedly
in the late 1940s. By the mid-1960s the expected growth in the metropolitan Phoenix and Tucson areas, coupled with the already large

Arizona water use by sector, 1994

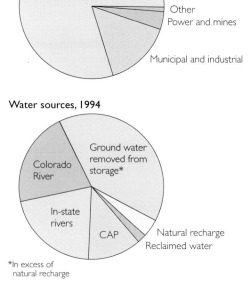

Water sources, 1994

*In excess of
natural recharge

(Arizona Department of Water Resources,
accessed July 27, 1999)

Agriculture in Arizona requires
intensive irrigation.

(U.S. Bureau of Reclamation)

ground-water-level declines and worsening subsidence problems, prompted Arizona water officials to push for and receive congressional approval for the CAP. Since then, growth in the metropolitan areas has exceeded expectations, and municipal-industrial and domestic water use presently accounts for nearly 20 percent of Arizona's water demand.

Subsidence follows water-level declines

Subsidence first became apparent during the 1940s in several alluvial basins in southern Arizona where large quantities of ground water were being pumped to irrigate crops. By 1950, earth fissures began forming around the margins of some of the subsiding basins. The areas affected then and subsequently include metropolitan Phoenix in Maricopa County and Tucson in Pima County, as well as important agricultural regions in Pinal and Maricopa Counties near Apache Junction, Chandler Heights, Stanfield, and in the Picacho Basin; in Cochise County near Willcox and Bowie; and in La Paz County in the Harquahala Plain. By 1980 ground-water levels had declined at least 100 feet in each of these areas and between 300 and 500 feet in most of the areas.

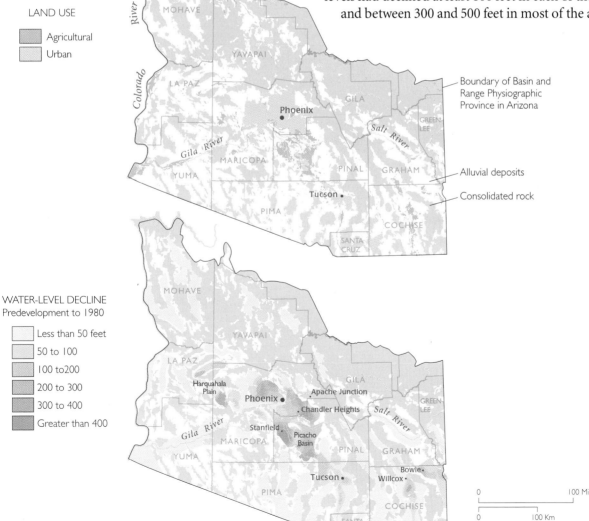

LAND USE

Agricultural

Urban

Boundary of Basin and
Range Physiographic
Province in Arizona

Alluvial deposits

Consolidated rock

WATER-LEVEL DECLINE
Predevelopment to 1980

Less than 50 feet

50 to 100

100 to 200

200 to 300

300 to 400

Greater than 400

(Anderson and others, 1992)

Central Arizona Project (CAP)
Delivering water to the interior basins

The primary purpose of The Central Arizona Project (CAP) is to help conserve the ground-water resources of Arizona by extending the supply of Colorado River water to interior basins in Arizona that are heavily dependent on the already depleted ground-water supplies. A body of legal doctrine collectively known as the "Law of the River" allots Arizona up to 2.85 million acre-feet of Colorado River water yearly, depending on availability. The Central Arizona Project was designed to deliver about 1.5 million acre-feet of Colorado River water per year to Maricopa, Pinal, and Pima Counties. Colorado River water fills the aqueduct at Lake Havasu near Parker and flows 336 miles to the San Xavier Indian Reservation southeast of Tucson, with the aid of pumping plants and pumping-stations with lifts that total about 3,000 feet. Of the more than 80 major customers, 75 percent are municipal or industrial, 13 percent are irrigation districts, and about

(U.S. Bureau of Reclamation)

A segment of the CAP aqueduct snakes through the desert west of Phoenix.

12 percent are Native American communities. CAP water was first delivered to Phoenix in 1986 and to Tucson in 1992. Having a higher salinity than the natural ground-water supplies it augments, CAP water is generally used in three ways—direct treatment and delivery; treatment, blending and delivery; and spread in percolation basins to artificially recharge the aquifer systems. Before it is distributed as drinking water, CAP water is disinfected and generally "softened." Of the 1.5 million acre-feet annual capacity of the CAP, only about 1 million acre-feet were being directly utilized as of 1997. Much of the balance was used to augment natural aquifer-system recharge through artificial-recharge pilot projects, in order to store water for future use and mitigate water-level declines and limit subsidence.

Land subsidence was first verified in south-central Arizona in 1948 using repeat surveys of bench marks near Eloy (Robinson and Peterson, 1962). By the late 1960s, installation and monitoring of borehole extensometers at Eloy, Higley Road south of Mesa, and at Luke Air Force Base, as well as analysis of additional repeat surveys, indicated that land subsidence was occurring in several areas. The areas of greatest subsidence corresponded with the areas of greatest water-level decline (Schuman and Poland, 1970).

By 1977, nearly 625 square miles had subsided around Eloy, where as much as 12.5 feet of subsidence was measured; another 425 square miles had subsided around Stanfield, with a maximum sub-

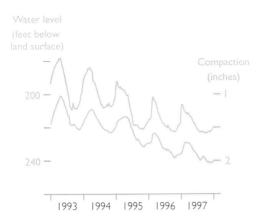

Water level
(feet below
land surface)

Compaction
(inches)

200 —   — 1

240 —   — 2

1993   1994   1995   1996   1997

Data from a borehole extensometer site in the Tucson Basin shows how compaction can
respond to water level changes.
Seasonal fluctuations are related to patterns of ground-
water pumping.

sidence of 11.8 feet (Laney and others, 1978). Near Queen Creek, an area of almost 230 square miles had subsided more than 3 feet. In northeast Phoenix, as much as 5 feet of subsidence was measured between 1962 and 1982. By contrast, in the Harquahala Plain, only about 0.6 feet of subsidence occurred in response to about 300 feet of water-level decline, whereas near Willcox, more than 5 feet of subsidence occurred in response to 200 feet of water-level decline (Holzer, 1980; Strange, 1983; Schumann and Cripe, 1986). The relation between water-level decline and subsidence varies between and within basins because of differences in the aggregate thickness and compressibility of susceptible sediments.

By 1992, ground-water level declines of more than 300 feet had caused aquifer-system compaction and land subsidence of as much as 18 feet on and near Luke Air Force Base, about 20 miles west of Phoenix. Associated earth fissures occur in three zones of differential subsidence on and near the base. Local flood hazards have greatly increased due to differential subsidence at Luke, which led to a flow reversal in a portion of the Dysart Drain, an engineered flood

Subsidence has occurred in basins with large water-level declines, but the relation between
the magnitude of water-level
decline and subsidence varies
between and within basins.
Representative profiles show
that subsidence is greater near
the center of basins, where the
aggregate thickness of fine-
grained sediments is generally
greater.

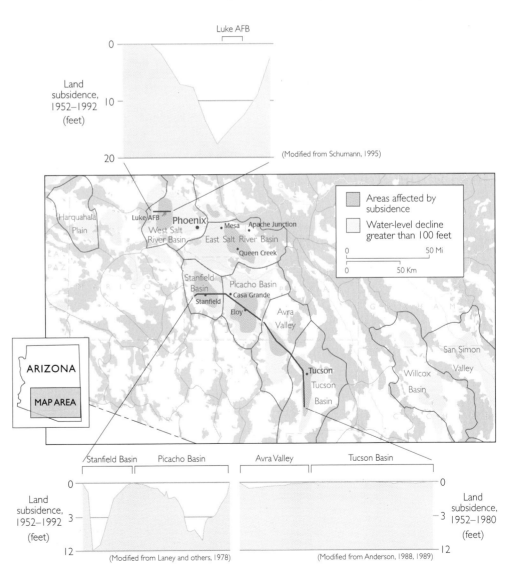

Fissures tend to develop near the margins of subsiding basins.

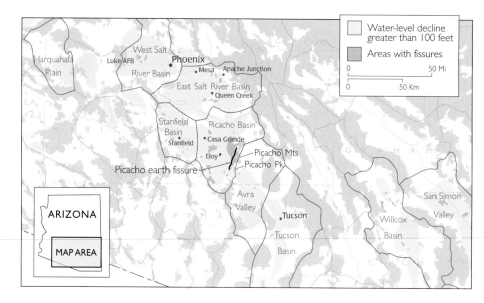

conveyance. On September 20, 1992, surface runoff from a rainstorm of 4 inches closed the base for 3 days. The sluggish Dysart Drain spilled over, flooding the base runways along with more than 100 houses and resulted in about $3 million in damage (Schumann, 1995).

EARTH FISSURES ARE COMMON IN MANY BASINS

Some of the most spectacular examples of subsidence-related earth fissures occur in south-central Arizona. Earth fissures are the dominant mode of ground failure related to subsidence in alluvial-valley sediments in Arizona and are typically long linear cracks at the land surface with little or no vertical offset. The temporal and spatial correlation of earth fissures with ground-water-level de-

Fissures have vertical sides, and typically first appear following severe rainstorms. Opening or movement is rarely more than 1 inch in any particular episode, although erosion and collapse of the sides during the initial episode may leave a fissure gully more than 10 feet wide, 30 feet deep, and hundreds of feet long. The apparent 1-foot width of the fissure that opened on July 23, 1976, near the Picacho Mountains, is due to erosion, collapse, and disintegration of down-dropped blocks. Several blocks remain wedged about 1 foot below land surface.

In another fissure that opened July 23, 1976, near the Picacho Mountains, an erosional gully 6 feet wide, 5 feet deep, and 20 feet long was cut in less than 16 hours. The head-cut gully developed perpendicular to the fissure in a wash on its upstream side. In subsequent storms, both the head-cut gully in the wash and the fissure were widened, deepened, and lengthened. It may take years or decades before a wash again carries water or sediment past a fissure that has cut across it.

clines indicates that many of the earth fissures are induced, and are related to ground-water pumpage. More than 50 fissure areas had been mapped in Arizona prior to 1980 (Laney and others, 1978).

Most fissures occur near the margins of alluvial basins or near exposed or shallow buried bedrock in regions where differential land subsidence has occurred. They tend to be concentrated where the thickness of the alluvium changes markedly. In a very early stage, fissures can appear as hairline cracks less than 0.02-inch wide interspersed with lines of sink-like depressions resembling rodent holes. When they first open, fissures are usually narrow vertical cracks less than about 1-inch wide and up to several hundred feet long. They

Fissure formation

Several theories explain the mechanism of fissure formation

Several mechanisms have been proposed for earth fissures, the most widely accepted of which is differential compaction. As ground-water levels decline in unconsolidated alluvial basins, less compaction and subsidence occurs in the thinner alluvium near the margin of the basin than in the thicker alluvium near the deeper, central part of the basin. The tension that results from the differential compaction stretches the overlying sediment until it fails as a fissure.

Differential compaction

As the land surface subsides, alluvium stretches and eventually fails, generally in a region of abrupt change in alluvium thickness.

Fissures are concentrated in areas where the thickness of the alluvium changes, such as near the margin of basins or where bedrock is near the surface.

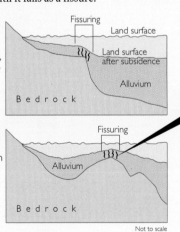

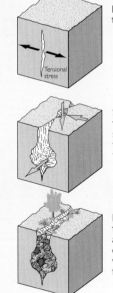

Lateral stresses induce tension cracking.

Surface water infiltrates, dissolving the natural cement bonding the soil, connecting hairline cracks, and further eroding and enlarging the fissure.

Fissure progressively enlarges, capturing surface runoff, sediment, and debris. Eventually vegetation establishes itself, creating a line of vegetation along the trace of the fissure.

OTHER POSSIBLE MECHANISMS

Horizontal seepage stresses and rotation of a rigid slab over an incompressible edge are other mechanisms that have been suggested. The observation that new fissures have formed between existing fissures and the mountain front argues against these two hypotheses. Hydrocompaction, or collapse of low-density soils upon complete wetting, and increased soil-moisture tension have also been suggested as possible mechanisms. Hydrocompaction in fact did occur during construction of sections of the CAP Aqueduct between the Picacho Mountains and Marana.

Other proposed mechanisms include piping erosion, soil rupture during earthquakes, renewed faulting, collapse of caverns or mines, oxidation of organic soils, and diapirism. Piping (subsurface soil erosion) along the trace of a fissure certainly plays a part in the opening, progressive enlargement and subsequent development of fissure gullies.

(Eaton and others, 1972; Carpenter, 1993)

Discovering Arizona's early fissures

Two fissures, two scientists, and their one discovery

On September 12, 1927, Professor R.J. Leonard from the University of Arizona visited and photographed an earth fissure south of the town of Picacho that was observed following a severe thunderstorm. After considering several possible causes for the fissure, Leonard tentatively concluded that an earthquake which had occurred on September 11, 1927, 170 miles from Tucson, caused the fissure by triggering the release of preexisting, accumulated strain. Leonard, a mining engineer, was probably influenced by his knowledge of the occurrence of unusual cracks at the El Tiro Mine near Silver Bell, Arizona, about 20 miles to the south (Leonard, 1929).

Two months later on November 13, 1927, Professor A.E. Douglas, also from the University of Arizona, visited and photographed what he probably thought was the same fissure that Leonard had photographed. In fact, it was not. The mountain skyline on Douglas's photographs lines up from a viewpoint about 1 mile to the southwest of Leonard's viewpoint. Leonard and Douglas discovered two separate earth fissures, and it was Douglas's photo that captured the precursor to the present-day Picacho earth fissure (Carpenter, 1993).

These early discoveries of multiple earth fissures at a time when ground-water withdrawals were just beginning raise some doubts about their origin. Although there is little doubt that ground-water-level declines since the 1940s have caused earth fissures, the cause of the Leonard and Douglas fissures remains a mystery.

Leonard's fissure

Douglas's view

(University of Arizona Tree Ring Laboratory photographs GEOL 27-2)

can progressively lengthen to thousands of feet. Apparent depths of fissures range from a few feet to more than 30 feet; the greatest recorded depth is 82 feet for a fissure on the northwest flank of Picacho Peak (Johnson, 1980). Fissure depths of more than 300 feet have been speculated based on various indirect measurements including horizontal movement, volume-balance calculations based on the volume of air space at the surface, and the amount of sediment transported into the fissures.

Widening of fissures by collapse and erosion results in fissure gullies (Laney and others, 1978) that may be 30-feet wide and 20-feet deep. No horizontal shear (strike-slip movement) has been detected at earth fissures, and very few fissures show any obvious vertical offset. However, fissures monitored by repeated leveling surveys commonly exhibit a vertical offset of a

A fissure moves with the seasonal fluctuation of water levels (data from the Picacho Basin).

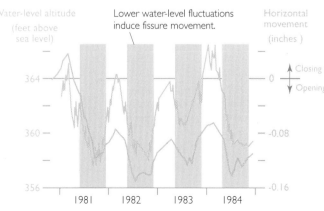

This aerial view taken in October 1967 shows the Picacho earth fissure as a single crack. A citrus grove is visible in the upper left.

By June 1989 the fissure had developed into a system of multiple parallel cracks. A fissure scarp developed as much as 2 feet of vertical offset, with the west or left side of the fissure (as pictured) down-dropped.

A lateral canal in the upper left skirts a citrus grove. This canal originates from the Central Arizona Project Aqueduct (not visible) at the base of the mountains in the background and crosses the fissure north of the citrus grove.

few inches. Two notable exceptions are the Picacho earth fissure, which has more than 2 feet of vertical offset at many places along its 10-mile length, and a fissure near Chandler Heights, which has about 1 foot of vertical offset.

The Picacho fissure is Arizona's most studied

The Picacho earth fissure, perhaps the most thoroughly investigated earth fissure (Holzer and others, 1979; Carpenter, 1993), began to creep vertically in 1961, forming a scarp. The scarp initially grew at a rate of more than 2 inches per year, before progressively slowing to about one-third inch per year by 1980 (Holzer, 1984). The observed opening and closing correlated with seasonal ground-water-level fluctuations from 1980 to 1984 (Carpenter, 1993). Surface deformation near the fissure indicated that formation of the vertical scarp was preceded by differential land subsidence and the formation of other earth fissures distributed over an approximately 1,000-foot-wide zone. Local geophysical and geologic surveys indicated that the Picacho earth fissure is associated with a preexisting high-angle, normal fault.

In the early 1950s Feth (1951) attributed formation of earth fissures west of the Picacho Mountains to differential compaction caused by ground-water-level decline in unconsolidated alluvium over the edge of a buried pediment or bedrock bench. He observed that fissures typically open during and after storms and potentially intercept large quantities of surface runoff. A decade later, the occurrence of subsidence-related fissures near Picacho, Chandler Heights, Luke Air Force Base, and Bowie was well known (Robinson and Peterson,

This fissure near the Picacho Mountains is undergoing erosional widening to become a fissure gully.

Another area experiencing subsidence-related earth fissures is near Casa Grande. This series of photographs shows how irrigation and pumping over a period of 22 years resulted in subsidence, surface depressions, and fissures possibly related to hydrocompaction.

A hydrograph from the well shown in the 1970 photo shows a sudden drop in water level after 1940.

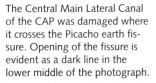

1949

UPLAND (consolidated rock)

BASIN FLOOR (alluvium)

Irrigated fields

Irrigated fields

1959

Depressions

Fissures

Depressions

1970

Fissure

Depressions

Fissure

Fissures

Depression

Well

0   0.5 Mile

The Central Main Lateral Canal of the CAP was damaged where it crosses the Picacho earth fissure. Opening of the fissure is evident as a dark line in the lower middle of the photograph.

1962). Subsidence-related earth fissures also have occurred in McMullen Valley (northwest of the Harquahala Plain), Avra Valley, the east Salt River Valley near Apache Junction, Willcox Basin (Schumann and Genauldi, 1986) and, as recently as 1997, in the Harquahala Plain (Al Ramsey, Arizona Department of Water Resources, written communication, 1998). Subsurface conditions beneath many subsidence-related earth fissures have been inferred principally from geophysical surveys and indicate that most occur above ridges or "steps" in the bedrock surface (Peterson, 1962; Holzer, 1984). In recent years, with introduction of CAP irrigation water, retirement of some farm lands, and the consequent recovery of water levels, earth fissures have apparently ceased to be active in some areas.

FISSURES CAN UNDERCUT AND DAMAGE INFRASTRUCTURE

Structures damaged by fissures include highways, railroads, sewers, canals, aqueducts, buildings, and flood-control dikes. The threat of damage from earth fissures forced a change in the proposed route of the CAP aqueduct. Erosionally enlarged fissure gullies present hazards to grazing livestock, farm workers, vehicles, hikers, and wildlife. Aquifer contamination may also occur as a result of ruptured pipelines, dumping of hazardous waste into fissures, and capture of surface runoff containing agricultural chemicals and other contaminants.

Where Interstate 10 crosses the Picacho earth fissure, more than 2 feet of vertical offset and several inches of horizontal opening have damaged the highway, requiring repeated pavement repairs. Where a natural gas pipeline crosses a fissure near the Picacho Mountains, erosional enlargement of the fissure left the pipeline exposed. The

Part of this fissure south of Apache Junction has been trenched and backfilled for a land bridge.

30-foot-wide hole was simply backfilled, but was repeatedly eroded for several years thereafter during summer and winter rains and had to be repeatedly refilled.

The CAP aqueduct and associated canals have been affected by earth fissures at several localities. Near Apache Junction, the U.S. Bureau of Reclamation installed vertical sheet piles on both sides of the CAP aqueduct in a fissure that undercuts the aqueduct. Soil beneath the aqueduct was compacted to reduce erosion. Erosional damage at this site and at another similarly treated site south of the Casa Grande Mountains has been minimal (Cathy Wellendorf, U.S. Bureau of Reclamation, written communication, 1988).

Engineering measures can also mitigate damage where fissures undercut roads. At Apache Junction, a trench was dug to a depth of about 30 feet, backfilled by about 10 feet of compacted fill, and then draped by a reinforced plastic grid, geotextile felt, and an impermeable membrane. The membrane was buried by additional compacted fill. This treatment protects the road from subsurface erosion by enhancing its structural strength and by restricting the upward flow of water from the fissure into the land bridge during flooding.

ARIZONA ACTS TO PROTECT THE AQUIFER SYSTEM

To ensure the future viability of the State's critical ground-water resources, the Arizona Groundwater Management Act was passed in 1980. This innovative law has three primary goals: (1) to control the severe overdraft of depleted aquifer systems, (2) to provide a means for allocating the limited ground-water resources among competing demands and effectively meet the changing needs of the State, and (3) to augment Arizona's ground-water resource through development of additional water supplies. The Act recognized ground water in Arizona as a public resource that must be managed for the benefit of everyone, and in 1986 was named one of the Nation's ten most innovative programs in State and local government by the Ford Foundation.

Based upon recommendations of the Groundwater Management Study Commission, which included representatives from cities and

A natural-gas pipeline undercut by an earth fissure was exposed through erosional widening of the fissure. The pipeline was evacuated and cut to determine the stresses on it. Tension was evident, but no shear.

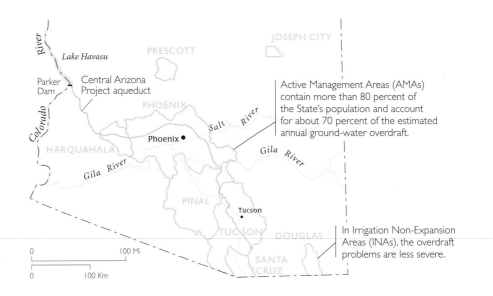

Active Management Areas (AMAs) contain more than 80 percent of the State's population and account for about 70 percent of the estimated annual ground-water overdraft.

In Irrigation Non-Expansion Areas (INAs), the overdraft problems are less severe.

For more information concerning the Arizona Groundwater Management Act, visit the Arizona Department of Water Resources web site at **http://www.adwr.state.az.us/**

towns, Native American communities, and mining, agricultural, and electric utilities industries, the Act focuses on limiting ground-water-level declines. Although it specifically mentions subsidence only three times, measures that limit ground-water-level declines will ultimately help to control compaction of the aquifer system and land subsidence. The Act provides for two levels of water management to respond to geographic regions where ground-water overdraft is a problem. Active Management Areas (AMAs) are designated where problems are most severe and Irrigation Non-Expansion Areas (INAs) are designated where problems are least severe. The Act established the Arizona Department of Water Resources (ADWR) to administer the Act. The State Director of the ADWR can designate additional AMAs for several reasons, including land subsidence or fissuring that is endangering property or potential ground-water-storage capacity (Carpenter and Bradley, 1986). The Act includes these six key provisions:

1. A program of ground-water rights and permits.

2. Restriction on new agricultural irrigation within AMAs.

3. Water conservation and management plans for AMAs that constitute 5 consecutive and progressively more stringent phases implemented during the periods 1980–1990, 1990–2000, 2000–2010, 2010–2020, and 2020–2025.

4. Assured water supply for new growth in AMAs before land may be marketed to the public.

5. Metering of ground-water pumpage for designated wells in AMAs.

6. Annual reporting of ground-water pumpage and assessment of withdrawal fees for designated wells in AMAs.

The original four AMAs were Phoenix, Pinal, Prescott, and Tucson. Subsequently, the Santa Cruz AMA was created by separation from the Tucson AMA in 1994. The two original INAs were Douglas and Joseph City, followed by Harquahala in 1982. The AMAs contain

A section of the Central Arizona Project passes through Apache Junction.

(U.S. Bureau of Reclamation)

more than 80 percent of the State's population and account for about 70 percent of the estimated annual ground-water overdraft in the State.

In the Tucson and Phoenix AMAs, which include the large urban areas of the State, and in the Prescott AMA, the primary management goal is to achieve safe yield by January 1, 2025. The goal in the Pinal AMA, where a predominantly agricultural economy exists, is to extend the life of the agricultural economy for as long as feasible and to preserve water supplies for future nonagricultural uses. In the Santa Cruz AMA, where significant ground-water/surface-water, international, and riparian water issues exist, the goal is to maintain safe yield and prevent the long-term decline of local unconfined aquifers.

Increasingly stringent conservation measures are being implemented in each of the AMAs during the five management periods. Municipal conservation measures include reductions in per capita water use measured in gallons per capita per day (GPCD). The requirements apply to the water providers, who must achieve target GPCDs through water-use restrictions or incentive-based conservation programs. Conservation for irrigated agriculture is being achieved by prohibiting new ground-water-irrigated acreage and by reductions in ground-water allotment, based on the quantity of water needed to irrigate the crops historically grown in the particular farm unit. There are also programs for augmenting water supplies, including incentives for artificial recharge, for purchase and retirement of irrigation rights, and for levying fees of up to $2.00 per acre-foot (Carpenter and Bradley, 1986).

A SUBSIDENCE-MONITORING PLAN WAS ESTABLISHED

In 1983, the National Geodetic Survey, with advice from an interagency Land Subsidence Committee, created a subsidence-monitoring plan for the Governor of Arizona. The plan summarized known subsidence and recognized hazards caused by subsidence, differential subsidence, and earth fissures in Arizona. The objectives of the plan were (1) "Documentation of the location and magnitude of existing subsidence and subsidence-induced earth fissures;" and (2) "Development of procedures for estimating future subsidence as a function of water-level decline and defining probable areas of future fissure development." The plan proposed a central facility at a State agency for compilation and organization of leveling, compaction, gravity, and other geophysical and stratigraphic information. There were plans to coordinate the analysis of existing data, to produce estimates of future subsidence and earth-fissure development, and to identify observation requirements. Other provisions included (1) "[a]n initial observation program designed to obtain a limited amount of additional leveling data, gravity observations, compaction measurements, and horizontal strain determinations;" and (2) "[a] cooperative effort between State and Federal agencies to evaluate new measurement technologies which offer the potential of being faster and more cost effective than current methods of subsidence monitoring." Also included were proposals for directions in research, some initial monitoring plans, and an advisory committee to oversee the formation of the central data facility and provide continuing guidance. (Strange, 1983). The recommendations have been only partially implemented. The Arizona Geological Survey has a Center for Land Subsidence and Earth Fissure Information. The USGS, the Arizona Department of Water Resources, the City of Tucson, and Pima County maintain cooperative programs for monitoring subsidence using global positioning system (GPS) surveying, microgravity surveys, and borehole extensometers. The ADWR has also started its own program of GPS surveying and microgravity surveys in the Phoenix metropolitan area.

In 1997, 19 of 29 borehole extensometers installed in south-central Arizona to measure aquifer-system compaction were still in operation. In the early 1990s, water levels in the Tucson basin continued to decline by as much as 3 to 6 feet per year, and a small amount of subsidence, generally less than 0.2 inch per year, was occurring in some areas. During the same period, water levels in Avra Valley continued to decline by 3 feet per year, and some subsidence, generally less than 0.1 inch per year, was occurring in some areas (City of Tucson Water Department, 1995). In the Picacho Basin, despite water-level recoveries of as much as 150 feet, some areas continue to subside at rates of up to 0.3 inches per year, most likely due to residual compaction of slowly equilibrating aquitards.

RISING WATER LEVELS OFFER SOME HOPE FOR THE FUTURE

(U.S. Bureau of Reclamation)

Importation of CAP water for consumptive use and ground-water recharge, retirement of some farmlands, and water-conservation measures have resulted in cessation of water-level declines in many areas and the recovery of water levels in some areas. However, some basins are still experiencing subsidence, because much of the aquifer-system compaction has occurred in relatively thick aquitards. It can take decades or longer for fluid pressures to equilibrate between the aquifers and the full thickness of many of these thick aquitards. For this reason, both subsidence and its abatement have lagged pumping and recharge. A glimmer of hope exists from data at the borehole extensometer near Eloy, where water levels have recovered more than 150 feet and compaction has decreased markedly.

PART II

Drainage of Organic Soils

Sacramento-San Joaquin Delta

Florida Everglades

Cultivated peat soils in the
Sacramento-San Joaquin Delta

(California Department of Water Resources)

I n the U.S. system of soil taxonomy, organic soils or histosols
are one of 10 soil orders. They are formally defined as having
more than 50 percent organic matter in the upper 30 inches, but
may be of lesser thickness if they overlie fragmental rock permeated
by organic remains. Organic soil is commonly termed "peat," if
fibrous plant remains are still visible, or "muck" where plant remains
are more fully decomposed. Other common names for accumulations of organic soil include "bog," "fen," "moor," and "muskeg."

Organic soils generally form in wetland areas where plant litter
(roots, stems, leaves) accumulates faster than it can fully decompose.
Fibrous peats typically include the remains of sedges and reeds that
grew in shallow water. "Woody" peats form in swamp forests. In
northerly latitudes with cool, moist climates, many peats are composed mainly of sphagnum moss and associated species. The total
area of organic soils in the United States is about 80,000 square
miles, about half of which is "moss peat" located in Alaska (Lucas,
1982). About 70 percent of the organic-soil area in the contiguous 48
States occurs in northerly, formerly glaciated areas, where moss
peats are also common (Stephens and others, 1984).

Most organic soils occur in
the northern contiguous
48 States and Alaska.

Land subsidence invariably occurs when organic soils are drained for agriculture or other purposes. There are a number of causes, including compaction, desiccation, erosion by wind and water, and, in some cases, prescribed or accidental burning. The effects of compaction and desiccation after initial draining can be dramatic, because organic soils have extremely low density and high porosity or saturated water content (up to 80 to 90 percent).

DRAINED ORGANIC SOILS WILL LITERALLY DISAPPEAR

The most important cause of organic-soil subsidence, however, is a process commonly termed "oxidation." The balance between accumulation and decomposition of organic material shifts dramatically when peat wetlands are drained. Under undrained conditions, anaerobic microbial decomposition of plant litter—that is, decomposition in the absence of free oxygen—cannot keep pace with the rate of accumulation. One reason is that lignin, an important cell-wall component of all vascular plants, is much more vulnerable to decomposition under aerobic conditions. Oxidation under aerobic conditions converts the organic carbon in the plant tissue to carbon dioxide gas and water. Aerobic decomposition under drained conditions is much more efficient.

The biochemical origin of much organic-soil subsidence was established by 1930 through laboratory experiments with Florida peat that balanced the loss of dry soil weight with rates of carbon-dioxide production (Waksman and Stevens, 1929; Waksman and Purvis, 1932). This early laboratory work also suggested optimal temperature ranges and moisture contents for microbial decomposition. Later field studies and observations have confirmed "oxidation" as the dominant subsidence process in many instances. For example, in the Florida Everglades, sod fields and residential areas—where causal mechanisms such as erosion, burning, and compaction are minimized or absent—have sunk as rapidly as the cultivated land (Stephens and others, 1984). It is believed that oxidation-related soil loss can be halted only by complete resaturation of the soil or complete consumption of its organic carbon content (Wosten and others, 1997).

Whereas natural rates of accumulation of organic soil are on the order of a few inches per 100 years, the rate of loss of drained organic soil can be 100 times greater, up to a few inches per year in extreme cases. Thus, deposits that have accumulated over many millennia can disappear over time scales that are very relevant to human activity.

SOME ORGANIC SOILS CAN BE CULTIVATED FOR CENTURIES

Human experience with subsiding organic soils dates back nearly 1,000 years in The Netherlands and several hundred years in the English fen country. The old polders in the western Netherlands were reclaimed for agriculture between the 9th and 14th centuries,

Evidence of subsidence in the Everglades is shown on a concrete marker that has been driven through the organic soil into the underlying limestone substrate.

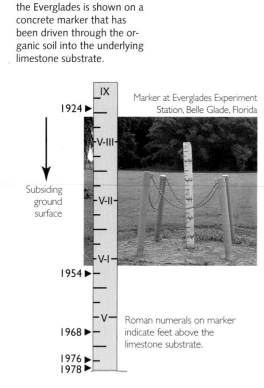

Marker at Everglades Experiment Station, Belle Glade, Florida

Roman numerals on marker indicate feet above the limestone substrate.

Subsiding ground surface

and by the 16th century the land had subsided to such an extent that windmills were needed to discharge water artificially to the sea (Shothorst, 1977). Because ground-water levels beneath the polders were still relatively high, the rate of subsidence was relatively low— less than 5 feet total, or 0.06 inches per year, over a roughly 1,000-year period in which progressively more sophisticated drainage systems were developed (Nieuwenhuis and Schokking, 1997). Greatly improved drainage in the 20th century increased the thickness of the drained zone above the water table. As a result, subsidence rates rose to about 0.2 inches per year between the late 1920s and late 1960s, and current rates are more than 0.3 inches per year.

The organic-soil subsidence rates in The Netherlands are still unusually low in a global context. This is due in part to the relatively cool climate, where temperatures are generally below the optimal range for microbial decomposition, and in part to a thin layer of marine clay that caps much of the peat. Larger average rates have been observed elsewhere: up to 3 inches per year over the last 100 years in the Sacramento-San Joaquin Delta, California; about 1 inch per year over the past 100 years in the English fens; and about 1 inch per year for the last 70 years in the Florida Everglades.

Both in the English fens and the Everglades, long-term subsidence rates have been monitored using stone or concrete columns driven into the underlying solid substrate. The history of both areas has been marked by alternate cycles of improved drainage followed by accelerated subsidence and, consequently, inadequate drainage (Stephens and others, 1984), so that the achievements of one generation become the problems of the next (Darby, 1956).

Long-term subsidence rates in the Everglades show cycles. Subsidence slows during periods of poor drainage and accelerates when pumps are installed to improve drainage.

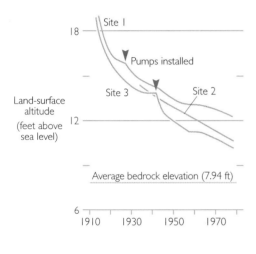

(Stephens and others, 1984)

SACRAMENTO-SAN JOAQUIN DELTA

The sinking heart of the state

CALIFORNIA

Sacramento

Sacramento River

Cosumnes R.

D E L T A

Chipps Island

Stockton

Mt. Diablo   San Joaquin River

0   15 Miles

0   15 Kilometers

S.E. Ingebritsen
U.S. Geological Survey,
Menlo Park, California

Marti E. Ikehara
National Geodetic Survey,
Sacramento, California

The Sacramento-San Joaquin Delta of California was once a great tidal freshwater marsh. It is blanketed by peat and peaty alluvium deposited where streams originating in the Sierra Nevada, Coast Ranges, and South Cascade Range enter San Francisco Bay.

In the late 1800s levees were built along the stream channels and the land thus protected from flooding was drained, cleared, and planted. Although the Delta is now an exceptionally rich agricultural area (over $500 million crop value as of 1993), its unique value is as a source of freshwater for the rest of the State. It is the heart of a massive north-to-south water-delivery system. Much of this water is pumped southward for use in the San Joaquin Valley and elsewhere in central and southern California.

The leveed tracts and islands help to protect water-export facilities in the southern Delta from saltwater intrusion by displacing water and maintaining favorable freshwater gradients. However, ongoing subsidence behind the levees increases stresses on the levee system, making it less stable, and thus threatens to degrade water quality in the massive north-to-south water-transfer system. Most subsidence in the Delta is caused by oxidation of organic carbon in peat soils.

THE DELTA MARSHES TEEMED WITH WILDLIFE

When Spanish explorers first viewed the Delta from Mount Diablo in 1772, the Sacramento and San Joaquin Rivers were in flood, and they mistook it for a great inland sea. In fact, the prehistoric Delta consisted largely of "tule" (bulrush) and reed marshes that were periodically submerged, with narrow bands of riparian forest on the natural levees along major stream channels. Exceptionally abundant fish and game supported a large

(The Nature Conservancy)

Native American population. When the Spanish first set foot in the Delta, they found the deer and tule elk trails to be so broad and extensive that they first supposed that the area was occupied by cattle. Similarly, American soldiers exploring the Delta in the 1840s found waterfowl to be so abundant and tame that they were mistaken for domestic fowl. The Native Americans were also able to harvest abundant local shellfish and the salmon that migrate through the Delta en route to spawning grounds in streams of the Sierra Nevada and southern Cascades.

Trappers from the Hudson Bay Company and elsewhere visited the Delta periodically between 1827 and 1849, drawn by the initially abundant beaver and river otter. By the beginning of the California Gold Rush in 1849, the Native American population of the Delta had been nearly destroyed by intermittent warfare with the Spanish and Mexicans and great epidemics of malaria (?) and cholera (1833) and smallpox (1839) (Dillon, 1982). Shortly after the Gold Rush, a great effort to control and drain the Delta for agriculture began. Levees were built along the stream channels, and the land thus protected from flooding was drained, cleared, and planted. The results of such reclamation seemed miraculous—in a letter to a friend, early settler George McKinney reported cabbages weighing 53 pounds per head and potatoes 33 inches in circumference (Dillon, 1982).

Agriculture and water now dominate the landscape

Today, the Delta is largely devoted to agriculture, and includes about 55 islands or tracts that are imperfectly protected from flooding by over 1,000 miles of levees. Many of the islands in the central Delta are 10 to nearly 25 feet below sea level because of land subsidence associated with drainage for agriculture. There are also numerous smaller, unleveed islands that remain near sea level. Remnants of the natural tule marsh are found on the unleveed "channel" or "tule" islands and along sloughs and rivers. The strips of natural riparian forest have nearly vanished, except on some of the larger channel islands, but relicts can be viewed at the Nature Conservancy's Cosumnes River Preserve in the northeastern Delta.

Although the Delta is an exceptionally productive agricultural area, its unique value to the rest of the State is as a source of freshwater. The Delta receives runoff from about 40 percent of the land area of California and about 50 percent of California's total streamflow. It is the heart of a massive north-to-south water-delivery system whose giant engineered arterials transport water southward. State and Federal contracts call for export of up to 7.5 million acre-feet per year from two huge pumping stations in the southern Delta near the Clifton Court Forebay (California Department of Water Resources, 1993). About 83 percent of this water is used for agriculture and the remainder for various urban uses in central and southern California. Two-thirds of California's population (more than 20 million people) gets at least part of its drinking water from the Delta (Delta Protection Commission, 1995).

The tule marshes of the Delta once teemed with migratory birds and fish.

(The Nature Conservancy)

Delta waterways pass through fertile farmland.

(California Department of Water Resources)

The Delta soils are composed of mineral sediments delivered by the rivers and of peat derived from decaying marsh vegetation. The peat began accumulating about 7,000 years ago and, prior to settlement, accumulated at a rate just sufficient to keep up with the average postglacial sea-level rise of about 0.08 inches per year (Atwater, 1980). The total thickness of peat was as large as 60 feet in the extreme western areas. The mineral sediments are more abundant on the periphery of the Delta and near the natural waterways, whereas the peat soils are thickest in former backwaters away from the natural channels—that is, towards the centers of many of the current islands.

The waterways of the entire Delta are subject to tidal action—tidal surges from San Francisco Bay are observed 5 hours later along the Cosumnes River in the eastern Delta. The position of the interface between the saline waters of the Bay and the freshwaters of the Delta depends upon the tidal cycle and the flow of freshwater through the Delta. Before major dams were built on rivers in the Delta watershed, the salinity interface migrated as far upstream as Courtland along the Sacramento River (California Department of Water Resources, 1993). Today, releases of freshwater from dams far upstream help reduce landward migration of the salinity interface during the summer months. A complicated formula agreed upon by all relevant parties attempts to maintain the two parts per thousand salinity interface near Chipps Island at the western edge of the Delta.

RECLAMATION FOR AGRICULTURE LED TO SUBSIDENCE

Sustained, large-scale agricultural development in the Delta first required levee-building to prevent frequent flooding. The levee-surrounded marshland tracts then had to be drained, cleared of tules, and tilled. The labor force for the initial levee-building effort consisted mainly of Chinese immigrants who arrived in large numbers upon completion of the Transcontinental Railroad in 1869. Between 1860 and 1880, workers using hand tools reclaimed about 140 square miles of Delta land for agriculture. The Chinese labor force was paid about a dollar per day, or at a piecework rate of 13 cents per cubic yard of material moved. After about 1880 the clamshell dredge, still in use today, became the dominant reclamation tool.

Chinese laborers built many of the early levees in the Delta.

(Overland Monthly, 1896)

A clamshell dredge operates near Sherman Island, ca. 1907.

(National Maritime Museum, San Francisco)

Levees and drainage systems were largely complete by 1930, and the Delta had taken on its current appearance, with most of its 1,150-square-mile area reclaimed for agricultural use (Thompson, 1957).

Reclamation and agriculture have led to subsidence of the land surface on the developed islands in the central and western Delta at long-term average rates of 1 to 3 inches per year (Rojstaczer and others, 1991; Rojstaczer and Deverel, 1993). Islands that were originally near sea level are now well below sea level, and large areas of many islands are now more than 15 feet below sea level. The land-surface profile of many islands is somewhat saucer-shaped, because subsidence is greater in the thick peat soils near their interior than in the more mineral-rich soils near their perimeter. As subsidence progresses the levees themselves must be regularly maintained and periodically raised and strengthened to support the increasing stresses on the levees that result when the islands subside. Currently, they are maintained to a standard cross section at a height 1 foot above the estimated 100-year-flood elevation.

Water levels in the depressed islands are maintained 3 to 6 feet below the land surface by an extensive network of drainage ditches, and the accumulated agricultural drainage is pumped through or over the levees into stream channels. Without this drainage the islands would become waterlogged.

"Watch that first step!"

The land surface has subsided beneath a Delta house, 1950.

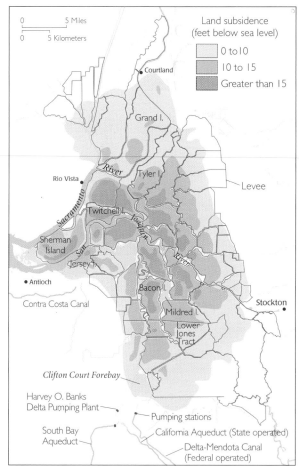

(California Department of Water Resources)

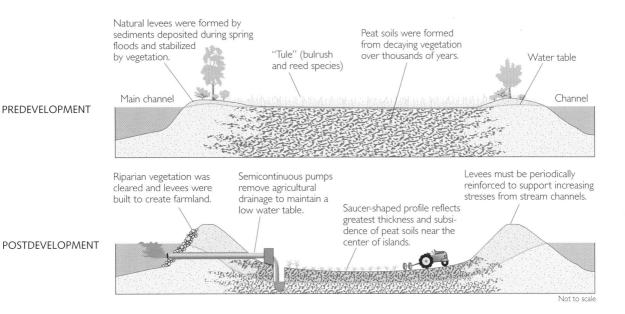

PREDEVELOPMENT

Natural levees were formed by sediments deposited during spring floods and stabilized by vegetation.

Main channel

"Tule" (bulrush and reed species)

Peat soils were formed from decaying vegetation over thousands of years.

Water table

Channel

POSTDEVELOPMENT

Riparian vegetation was cleared and levees were built to create farmland.

Semicontinuous pumps remove agricultural drainage to maintain a low water table.

Saucer-shaped profile reflects greatest thickness and subsidence of peat soils near the center of islands.

Levees must be periodically reinforced to support increasing stresses from stream channels.

Not to scale

Decomposing peat soils are the main cause of subsidence

The dominant cause of land subsidence in the Delta is decomposition of organic carbon in the peat soils. Under natural waterlogged conditions, the soil was anaerobic (oxygen-poor), and organic carbon accumulated faster than it could decompose. Drainage for agriculture led to aerobic (oxygen-rich) conditions. Under aerobic conditions microbial activity oxidizes the carbon in the peat soil quite rapidly. Most of the carbon loss from the soil occurs as a flux of carbon-dioxide gas to the atmosphere.

Pumps, such as these on Twitchell Island, remove agricultural drainage while maintaining the water table at a level low enough to sustain agriculture.

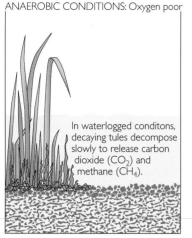

ANAEROBIC CONDITIONS: Oxygen poor

In waterlogged conditons, decaying tules decompose slowly to release carbon dioxide (CO_2) and methane (CH_4).

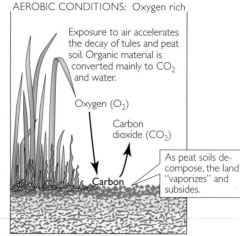

AEROBIC CONDITIONS: Oxygen rich

Exposure to air accelerates the decay of tules and peat soil. Organic material is converted mainly to CO_2 and water.

Oxygen (O_2)

Carbon dioxide (CO_2)

Carbon

As peat soils decompose, the land "vaporizes" and subsides.

Scientists resolve subsidence mechanisms

There has been some debate as to the causes and mechanisms of subsidence in the Delta. Possible causes include deep-seated compaction related to the removal of subsurface fluids (oil, gas, and water) and the near-surface oxidation and mass wasting of organic soils. This debate seems to have been resolved in favor of the carbon oxidation/gas flux hypothesis. Extensometer measurements have shown that deep-seated subsidence due to natural-gas production and ground-water withdrawal is minimal. Further, pockets of unreclaimed marshland on channel islands remain at sea level. Age-dating of sediment cores from these islands indicates low sedimentation rates and, by inference, minimal subsidence in unreclaimed areas (Rojstaczer and others, 1991). These studies made it clear that Delta subsidence is a near-surface process, but did not establish how the carbon loss takes place. Further studies by the USGS, in cooperation with the California Department of Water Resources, resolved this issue by simultaneously measuring subsidence and carbon fluxes at several sites (Deverel and Rojstaczer, 1996). The increased gaseous flux of carbon dioxide was sufficient to explain most of the carbon loss and measured subsidence, whereas the dissolved organic carbon (DOC) pumped from the islands in agricultural drainage could account for only about 1 percent of the carbon loss.

The USGS experiments also showed that rates of carbon-dioxide production increase with increasing temperature and decrease with increasing soil moisture. These results are consistent with field and laboratory measurements from the Florida Everglades, where subsidence is occurring by the same mechanism, albeit at a smaller rate of about 1 inch per year.

The rate of subsidence has decreased

The best evidence for long-term rates of subsidence comes from two sources—measurements of the exposure of transmission-line

These transmission towers on Sherman Island show evidence of subsidence

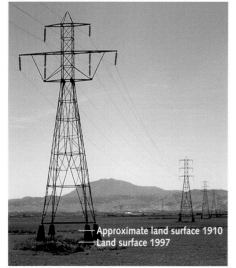

Approximate land surface 1910
Land surface 1997

How to slow or reverse subsidence
Scientists look for answers with controlled experiments

Investigations on various islands in the Sacramento-San Joaquin Delta have shown that microbial decomposition of organic-rich soils is causing the land to "vaporize" and disappear. Ongoing experiments at two sites on Twitchell Island in the western Delta focus on assessing the factors that affect the rate and timing of carbon-dioxide production.

At one of the Twitchell Island sites, the land surface is subjected to a variety of flooding scenarios in order to assess anaerobic and aerobic decomposition processes.

At the other site (not shown), which will be permanently flooded, the effects of vegetative cover on the potential for biomass accumulation will be assessed.

Tules will be planted on subsets of this site and will spread throughout the site. They will decompose relatively slowly under flooded conditions. It is anticipated that plant-litter accumulations will become peat-like material over time and eventually increase land-surface elevations measured relative to stable markers set in mineral soil beneath the peat.

FUTURE STRATEGIES

Possible long-term management strategies for various Delta islands include:

1. Shallow flooding to slow peat oxidation and reverse subsidence through biomass accumulation.

2. Shallow flooding combined with thin-layer mineral deposition (a possibly beneficial reuse of dredge material).

3. Continued agricultural use of areas with shallow peat and/or low organic-matter content, under the assumption that the maximum additional subsidence will not destabilize the levees.

4. Blending mineral soil with peat soil to decrease the rate of carbon dioxide (CO_2) release and allow continued agricultural use.

5. Addition of thick layers of mineral soil, possibly using controlled levee breaches or deposition of dredge material, to slow peat oxidation and raise land-surface elevation.

6. Deep flooding to create freshwater reservoirs.

These strategies may be implemented in a mosaic throughout the Delta that creates a substantial diversity of wildlife habitat—uplands, open water, shallow permanent wetlands, and seasonal wetlands.

foundations on Sherman and Jersey Islands in the western Delta and repeated leveling surveys on Mildred and Bacon Islands and Lower Jones Tract in the southern Delta (Weir, 1950; Rojstaczer and others, 1991). The transmission lines in the western Delta were installed in 1910 and 1952. They are founded on pylons driven down to a solid substrate, so that comparison of the original foundation exposure with the current exposure allows estimates of soil loss. The southern Delta transect was surveyed 21 times between 1922 and 1981; in 1983 further surveys were precluded when Mildred Island flooded. Both data sets indicate long-term average subsidence rates of 1 to 3 inches per year, but also suggest a decline in the rate of subsidence over time, probably due to a decreased proportion of readily decomposable organic carbon in the near surface (Rojstaczer and Deverel,1993). In fact, rates of elevation loss measured at three selected sites in 1990 to 1992 were less than 0.4 inches per year, consistent with the inferred slowing of subsidence (Deverel and Rojstaczer, 1996). However, all of these sites were near island edges, and likely underestimate the average island-wide elevation loss.

MANY MANAGEMENT ISSUES ARE RELATED TO SUBSIDENCE

The management issues raised by land subsidence range in scale from those faced by individual farmers to the possible global-scale

Living with possible levee failure
Approximately 1,100 miles of levees need to be maintained

Levee failure has been common in the Sacramento-San Joaquin Delta since reclamation began in the 1850s. Each of the islands and tracts in the Delta has flooded at least once, with several flooding repeatedly. About 100 levee failures have occurred since the early 1890s. Initially, most of the failures were caused by overtopping during periods of spring flooding. Although construction of upstream reservoirs since the 1940s has reduced the threat of overtopping, it has not reduced the incidence of levee failure.

Dredge material is used to reinforce levees.

Tyler Island levee was breached in a 1986 flood.

(California Department of Water Resources)

EARTHQUAKES

The Delta sits atop a blind fault system on the western edge of the Central Valley. Moderate earthquakes in 1892 near Vacaville and in 1983 near Coalinga demonstrate the seismic potential of this structural belt.

The increasing height of the levee system has prompted growing concern about the seismic stability of the levees. The concern is based on the proximity of faulting, the nature of the levee foundations, and the materials used to build the levees. Many levees consist of uncompacted weak local soils that may be unstable under seismic loading. The presence of sand and silt in the levees and their foundations indicates that liquefaction is also a possibility. Although no historic examples of seismically induced levee failure are known in the Delta, the modern levee network has not been subjected to strong shaking. Levees were either smaller or nonexistent in 1906 when the region was strongly shaken by the great San Francisco earthquake.

Areas with peat thickness over 10 feet have a great potential for continued subsidence.

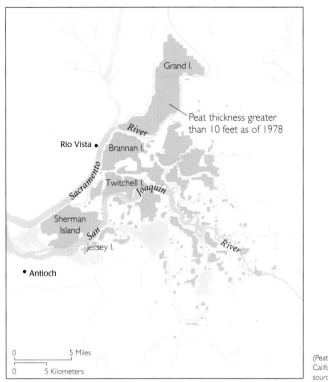

Peat thickness greater than 10 feet as of 1978

(Peat-thickness estimates are from the California Department of Water Resources, written communication, 1998)

issue posed by the carbon-dioxide flux, with its possible link to climate change. At the most local level, individual farmers or reclamation districts must maintain drainage networks on the islands and pump the agricultural drainage back into waterways. These costs increase gradually as subsidence progresses.

As subsidence continues, levees must be enlarged

The costs of levee construction and maintenance are borne by the State of California and the Federal government, as well as by local reclamation districts. These costs also increase as subsidence progresses, forcing levees to be built higher and stronger. In 1981 to 1986 the total amount spent on emergency levee repairs related to flooding was about $97 million, and in 1981 to 1991 the amount spent on routine levee maintenance was about $63 million (California Department of Water Resources, 1993). Thus the annual cost of repair and maintenance of Delta levees in the 1980s averaged about $20 million per year.

Subsidence could affect California's water system

Much larger costs might be incurred if land subsidence indirectly affects the north-to-south water-transfer system, which is predicated on acceptable water quality in the southern Delta. The western Delta islands, in particular, are believed to effectively inhibit the inland migration of the salinity interface between Bay and Delta. If these are flooded, the water available to the massive pumping facilities near the Clifton Court Forebay might become too saline to use.

The fertile soils of the Delta are vulnerable to flooding.

(California Department of Water Resources)

Sacramento-San Joaquin Delta
The heart of California's water systems

An artificial balance is maintained in the water exchanged between the Delta and the San Francisco Bay. Freshwater inflows regulated by upstream dams and diversions supply water to the Delta ecosystems and to farms and cities in central and southern California. Subsidence of Delta islands threatens the stability of island levees and the quality of Delta water. Delta levee failures would tip the water-exchange balance in favor of more saltwater intrusion, which can ruin the water for agriculture and domestic uses. Several

aqueducts would be affected. Any reductions in the supply of imported Delta water could force water purveyors in many parts of the State to meet water demand with groundwater supplies. And this, in turn, could renew land subsidence in Santa Clara and San Joaquin Valleys and exacerbate subsidence in the Antelope Valley and other areas currently reliant on imported Delta water supplies and prone to aquifer-system compaction.

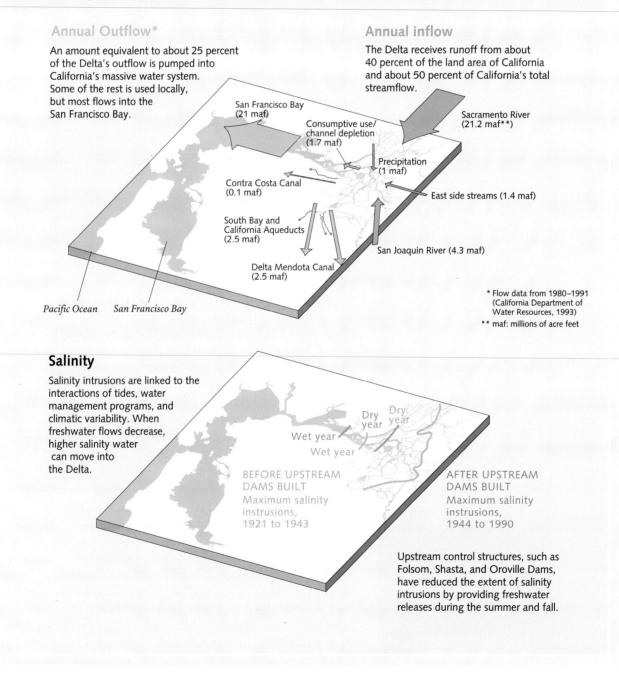

Annual Outflow*

An amount equivalent to about 25 percent of the Delta's outflow is pumped into California's massive water system. Some of the rest is used locally, but most flows into the San Francisco Bay.

Annual inflow

The Delta receives runoff from about 40 percent of the land area of California and about 50 percent of California's total streamflow.

San Francisco Bay (21 maf)

Consumptive use/ channel depletion (1.7 maf)

Sacramento River (21.2 maf**)

Precipitation (1 maf)

Contra Costa Canal (0.1 maf)

East side streams (1.4 maf)

South Bay and California Aqueducts (2.5 maf)

Delta Mendota Canal (2.5 maf)

San Joaquin River (4.3 maf)

Pacific Ocean   San Francisco Bay

* Flow data from 1980–1991 (California Department of Water Resources, 1993)

** maf: millions of acre feet

Salinity

Salinity intrusions are linked to the interactions of tides, water management programs, and climatic variability. When freshwater flows decrease, higher salinity water can move into the Delta.

Dry year
Dry year
Wet year
Wet year

BEFORE UPSTREAM DAMS BUILT
Maximum salinity instrusions, 1921 to 1943

AFTER UPSTREAM DAMS BUILT
Maximum salinity instrusions, 1944 to 1990

Upstream control structures, such as Folsom, Shasta, and Oroville Dams, have reduced the extent of salinity intrusions by providing freshwater releases during the summer and fall.

The Harvey O. Banks pumping plant at the southern edge of the Delta lifts water (lower right) into the California aqueduct (center left). The white towers on the upper left are wind turbines that generate electricity.

(California Department of Water Resources)

The timing of levee breaks and flooding is critical in this regard. Fortunately, most flooding occurs in winter and spring, when major saltwater intrusion is less likely. However, there are occasional levee failures under low-flow conditions. These can cause major short-term water-quality problems, even if the flooded areas are later reclaimed. During one island flooding under low-flow conditions, chloride levels reached 440 parts per million (ppm) at the Contra Costa Canal intake, well above the California standard for drinking water of 250 ppm (California Department of Water Resources, 1995).

The statewide water-transfer system in California is so interdependent that decreased water quality in the Delta might lead to accelerated subsidence in areas discussed elsewhere in this Circular. Both the Santa Clara and San Joaquin Valleys rely, in part, on imported water from the Delta to augment local supplies and thereby reduce local ground-water pumpage and arrest or slow subsidence. Degradation of the Delta source water could well lead to increased ground-water use, and renewed subsidence, in these and other areas in California.

Peat soil agriculture plays a minor role in climate change

The fact that most subsidence in the Delta, and in other drained wetlands, is caused by carbon oxidation suggests that such subsidence might affect atmospheric carbon-dioxide levels. The worldwide annual production of atmospheric carbon due to agricultural drainage of organic soils has been estimated to be as much as 6 percent of that produced by fossil fuel combustion (Tans and others, 1990). However, current rates of carbon-dioxide production in the Delta are likely to be significantly less than those caused by the initial agricultural expansion into virgin areas (Rojstaczer and Deverel, 1993). The gradual slowing of subsidence is associated with a declining rate of carbon-dioxide production.

THE FUTURE OF THE DELTA POSES MANY CHALLENGES

In cases where subsidence is due to aquifer-system compaction, it can often be slowed or arrested by careful water-use management. In cases where subsidence is due to peat oxidation, such as the Delta, it can be controlled only by major changes in land-use practice. In standard agricultural practice, the ultimate limiting factor is simply the total peat thickness; that is, the availability of organic carbon in the soil. In the Florida Everglades, the original peat thickness was less than 12 feet, and most of the potential subsidence has already been realized. In much of the cultivated area of the Delta, however, substantial thicknesses of peat remain, so that there is great potential for further subsidence.

Like the Everglades, the Delta is currently the subject of a major Federal-State restoration effort that includes attempts to improve wildlife habitat. These attempts have focused on the periphery of the Delta, avoiding the central areas with significant amounts of subsidence. As in the Everglades, much of the extensively subsided area is impractical to restore and will continue to be intensively managed.

As subsidence progresses, the levee system will become increasingly vulnerable to catastrophic failure during floods and earthquakes. The interrelated issues of Delta land subsidence, water quality, and wildlife habitat will continue to pose a major dilemma for California water managers.

This view of the Delta was taken looking westward with Mount Diablo on the horizon.

(California Department of Water Resources)

FLORIDA EVERGLADES

Subsidence threatens agriculture and complicates ecosystem restoration

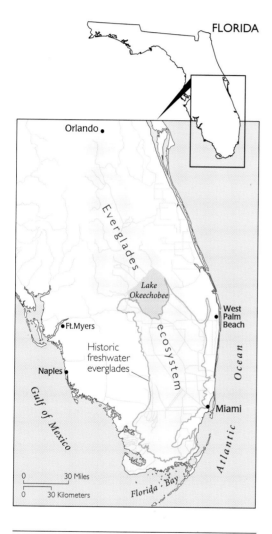

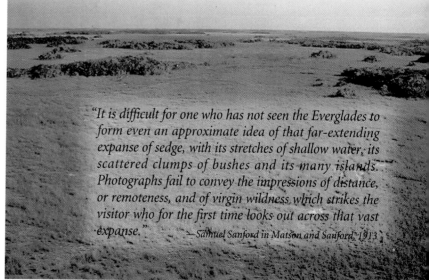

The Everglades ecosystem includes Lake Okeechobee and its tributary areas, as well as the roughly 40- to 50-mile-wide, 130-mile-long wetland mosaic that once extended continuously from Lake Okeechobee to the southern tip of the Florida peninsula at Florida Bay.

Since 1900 much of the Everglades has been drained for agriculture and urban development, so that today only 50 percent of the original wetlands remain. Water levels and patterns of water flow are largely controlled by an extensive system of levees and canals. The control system was constructed to achieve multiple objectives of flood control, land drainage, and water supply. More recently, water-management policies have also begun to address issues related to ecosystem restoration. Extensive land subsidence that has been caused by drainage and oxidation of peat soils will greatly complicate ecosystem restoration and also threatens the future of agriculture in the Everglades.

"It is difficult for one who has not seen the Everglades to form even an approximate idea of that far-extending expanse of sedge, with its stretches of shallow water, its scattered clumps of bushes and its many islands. Photographs fail to convey the impressions of distance, or remoteness, and of virgin wildness which strikes the visitor who for the first time looks out across that vast expanse." —Samuel Sanford in Matson and Sanford, 1913

S.E. Ingebritsen
U.S. Geological Survey, Menlo Park, California

Christopher McVoy
South Florida Water Management District, West Palm Beach, Florida

B. Glaz
U.S. Department of Agriculture, Agricultural Research Service, Canal Point, Florida

Winifred Park
South Florida Water Management District, West Palm Beach, Florida

The Everglades were formed in a limestone basin, which accumulated layers of peat and mud bathed by freshwater flows from Lake Okeechobee.

NATURAL FLOW PATTERNS (c.1900)

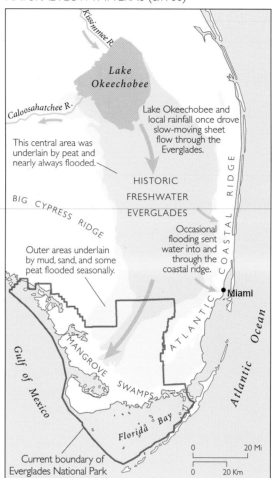

Kissimmee R.

Lake Okeechobee

Caloosahatchee R.

Lake Okeechobee and local rainfall once drove slow-moving sheet flow through the Everglades.

This central area was underlain by peat and nearly always flooded.

HISTORIC FRESHWATER EVERGLADES

BIG CYPRESS RIDGE

COASTAL RIDGE

Occasional flooding sent water into and through the coastal ridge.

Outer areas underlain by mud, sand, and some peat flooded seasonally.

Miami

MANGROVE SWAMPS

ATLANTIC

Gulf of Mexico

Florida Bay

Atlantic Ocean

Current boundary of Everglades National Park

0   20 Mi

0   20 Km

"The outline of this Florida end-of-land, within the Gulf of Mexico, the shallows of the Bay of Florida and the Gulf Stream, is like a long pointed spoon. That is the visible shape of the rock that holds up out of the surrounding sea water the long channel of the Everglades and their borders. The rock holds all the fresh water and the grass and all those other shapes and forms of air-loving life only a little way out of the salt water, as a full spoon lowered into a full cup holds two liquids separate, within that thread of rim."

—Marjorie Stoneman Douglas, 1947

The Everglades ecosystem has, in fact, been badly degraded, despite the establishment of Everglades National Park in the southern Everglades in 1947. Prominent symptoms of the ecosystem decline include an 80 percent reduction in wading bird populations since the 1930s (Ogden, 1994), the near-extinction of the Florida panther (Smith and Bass, 1994), invasions of exotic species (Bodle and others, 1994), and declining water quality in Florida Bay, which likely is due, at least in part, to decreased freshwater inflow (McIvor and others, 1994).

HISTORIC FLOWS WERE SEVERED

A thin rim of bedrock protects south Florida from the ocean. The limestone bedrock ridge that separates the Everglades from the Atlantic coast extends 20 feet or less above sea level. Under natural conditions all of southeast Florida, except for a 5- to 15-mile-wide strip along this bedrock ridge, was subject to annual floods. Much of the area was perennially inundated with freshwater. Water levels in Lake Okeechobee and local rainfall drove slow-moving sheet flow through the Everglades under topographic and hydraulic gra-

Everglades National Park was created in 1947.

Hoover dike (center) was built with digging spoils obtained from a navigable channel (foreground). Lake Okeechobee is at the top of photo.

dients of only about 2 inches per mile. Lake Okeechobee, which once overflowed its southern bank at water levels in the range of 20 to 21 feet above sea level, today is artificially maintained at about 13 to 16 feet above sea level by a dike system and canals to the Atlantic and Gulf coasts.

Early agriculturalists began the drying process

The first successful farming ventures in the Everglades began in about 1913, not on the sawgrass plain itself but on the slightly elevated natural levee south of Lake Okeechobee (Snyder and Davidson, 1994). Early efforts to clear, farm, and colonize the sawgrass area had little success, being plagued by flooding, winter freezes, and trace-nutrient deficiencies. (The soil beneath the sawgrass was later shown to be too low in copper to support most crops and livestock.)

In the 1920s the State of Florida established an Everglades Experiment Station in Belle Glade, and the U.S. Department of Agriculture established a Sugarcane Field Station in Canal Point. The combined efforts of these units gradually solved the plant- and livestock-pathology problems experienced by early farmers. However, the land was still subject to frequent, sometimes catastrophic inundation. The great hurricane of 1928 caused at least 2,000 fatalities and flooded the Everglades Experiment Station for several months.

The damage caused by the 1928 hurricane convinced the Federal government to fund construction of a permanent dike around the southern perimeter of Lake Okeechobee. This more secure protection from flooding cleared the way for intensive settlement of the Everglades. It also permanently severed the natural connection between the Everglades proper and its headwaters. For millennia, the Everglades had been fed by intermittent, diffuse overflow of the imperfect natural levee south of the Lake. Now, its primary water source, other than local rainfall, would be a system of artificial canals.

A network of dikes and canals controls water movement, providing optimum irrigation and drainage for sugar cane (left).

Further water-management efforts accelerated development

A comprehensive Federal-State water-management effort in the 1950s and 1960s was prompted by drought and widespread fires in 1944 to 1945 and renewed flooding in 1947 to 1948. The primary motivation was flood control and water supply for the growing urban areas along the Atlantic coast. The drying of the Everglades had clearly contributed to rapid saltwater intrusion in these urbanizing areas during the drought.

A regional flood-control district, the predecessor of today's South Florida Water Management District, was created by the State of Florida in 1949 to manage a coordinated water system. The urbanizing areas that extended west of the natural bedrock ridge were protected from flooding by a high levee known as the "eastern perimeter levee." Although it was originally built to protect and promote development of urbanizing areas along the southeastern

Water-control projects in the Everglades began in the early 1900s. After the fires and floods of the 1940s, much larger water-management projects were implemented.

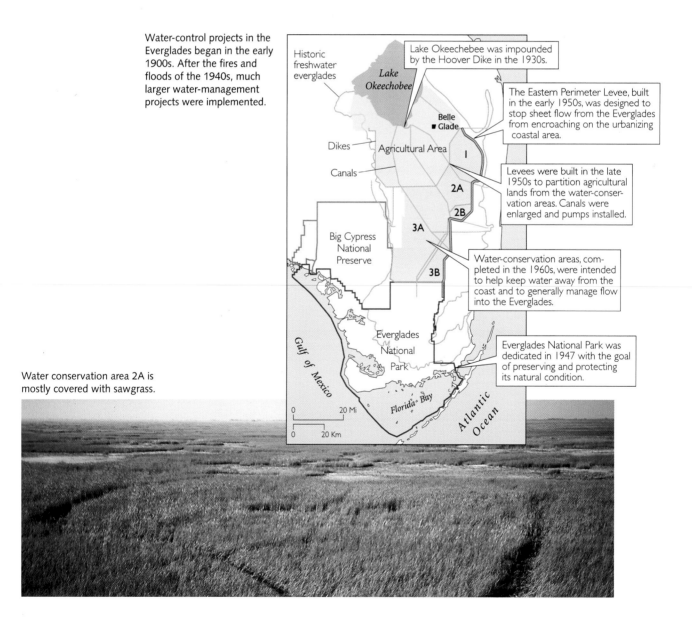

Lake Okeechebee was impounded by the Hoover Dike in the 1930s.

The Eastern Perimeter Levee, built in the early 1950s, was designed to stop sheet flow from the Everglades from encroaching on the urbanizing coastal area.

Levees were built in the late 1950s to partition agricultural lands from the water-conservation areas. Canals were enlarged and pumps installed.

Water-conservation areas, completed in the 1960s, were intended to help keep water away from the coast and to generally manage flow into the Everglades.

Everglades National Park was dedicated in 1947 with the goal of preserving and protecting its natural condition.

Water conservation area 2A is mostly covered with sawgrass.

NATURAL FLOW PATTERNS (ca. 1900)

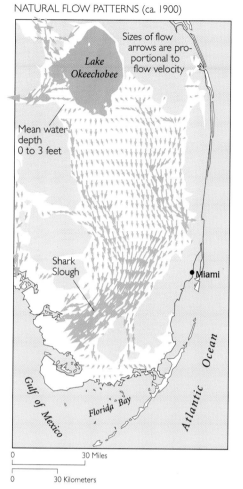

CURRENT FLOW PATTERNS (ca. 1990)

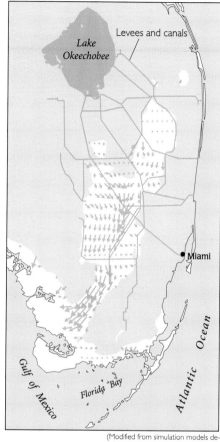

(Modified from simulation models developed and maintained by the South Florida Water Management District)

Water management has brought significant changes to natural overland flow patterns.

Under natural conditions surface water moved from Lake Okeechobee southward, then turned southwest through a constricted area called Shark Slough.

After canals and dikes were constructed for the agricultural and water-conservation areas, sheet flow practically disappeared from the northern Everglades and diminished to the south.

coast, this levee has, ironically, become the only effective barrier to more extensive urban development of the Everglades (Light and Dineen, 1994).

An area of thick peat soil south of Lake Okeechobee was designated the "Everglades agricultural area." Farther south, other areas of peat soils less suitable for agriculture were designated as "water-conservation areas." These areas are maintained in an undeveloped state, but a system of dikes and canals allows water levels to be manipulated to achieve management objectives that include flood control, water supply, and wildlife habitat.

During dry periods, the level of Lake Okeechobee drops as water is released to provide water to the agricultural area, to canals that maintain ground-water levels in urban areas along the Atlantic coast, and to Everglades National Park. At other times, drainage water pumped from the agricultural area is released into the water-conservation areas, providing needed water but also undesirable amounts of the nutrient phosphorus. In recent years, "best management practices" have helped reduce phosphorus loads from the agricultural area. The managed part of the remaining Everglades—approximately the northern two-thirds—now consists of a series of linked, impounded systems that are managed individually.

HISTORIC EVERGLADES VEGETATION (ca.1900)

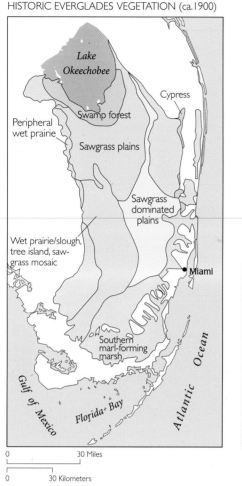

CURRENT EVERGLADES VEGETATION (ca.1990)

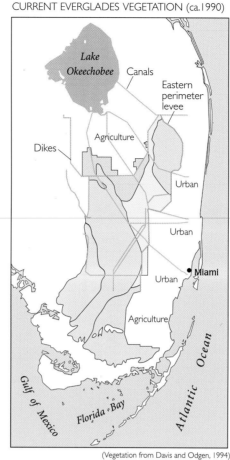

(Vegetation from Davis and Odgen, 1994)

Water management has also changed vegetation patterns. The construction of canals and levees and subsequent draining and development of the land has all but eliminated natural vegetation in the agricultural area and the region east of the eastern perimeter levee.

Land subsidence followed in the wake of development

With the addition of trace nutrients, the peat soil or "muck" beneath the sawgrass proved extremely productive. But the farmers also saw

"... the cushiony layer of dark muck shrink and oxidize under the burning sun as if it was consumed in thin, airy flames. As the canals and ditches were extended by the local drainage boards, and the peaty muck was dried out and cultivated, it shrank ... It is still shrinking. Every canal and ditch that drained it made a long deepening valley in the surrounding area. On the east and south the subsidence was so great that half that land [drains towards] the lowered lake."   —Marjorie Stoneman Douglas, 1947

In today's Everglades agricultural area, evidence of substantial land subsidence can readily be discerned from the relative elevations of the land surface, the drainage-canal system, and the lake, and from the elevation of older buildings that were built on piles extending to bedrock. Precise measurements are relatively rare, except at particular points or along a few infrequently revisited transects. However, the long-term average rate of subsidence is generally considered to have been between 1 and 1.2 inches per year (Stephens and Johnson, 1951; Shih and others, 1979; Stephens and others, 1984).

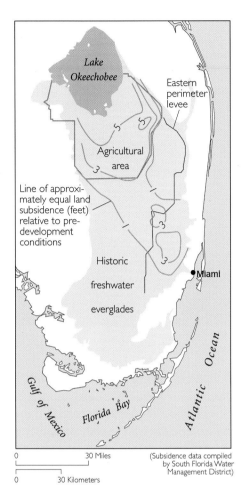

Line of approximately equal land subsidence (feet) relative to pre-development conditions

Historic freshwater everglades

Lake Okeechobee

Eastern perimeter levee

Agricultural area

Miami

Gulf of Mexico

Florida Bay

Atlantic Ocean

0   30 Miles

0   30 Kilometers

(Subsidence data compiled by South Florida Water Management District)

Subsidence is greater in areas that were intentionally drained for urban and agricultural uses.

In uncultivated areas of the Everglades, subsidence is less obvious but probably widespread. Subsidence is not caused by cultivation, but occurs wherever drainage desaturates peat soil. Early engineering efforts focused on drainage alone, and, as a result, much of the area became excessively drained during drought years. The "river of grass" often became a string of drying pools, and great fires swept the Everglades. The drying triggered subsidence, which was then exacerbated by widespread fires. The persistent peat fires sometimes continued smoldering for months before being extinguished by the next rainy season.

Conventional surveying has always been extremely difficult in the Everglades. Stable bedrock bench marks are nonexistent or very distant, the surficial material is soft and yielding, and access is difficult. Current best estimates suggest that there have been 3 to 9 feet of subsidence in the current Everglades agricultural area and that an equally large uncultivated area has experienced up to 3 feet of subsidence. Such elevation changes are tremendously significant to a near-sea-level wetlands system in which flow is driven by less than 20 feet of total relief.

The current management infrastructure and policies have abated land subsidence in undrained areas of the historic Everglades to some extent, although comparison of recent soil-depth measure-

This building at the Everglades Experiment Station was originally constructed at the land surface; latticework and stairs were added after substantial land subsidence.

A sugar mill outside Belle Glade is surrounded by sugar cane fields. Note the dark peat soils in the lower photograph.

ments by the U.S. Environmental Protection Agency (Scheidt, US EPA, written communication 1997) with 1940s estimates of peat thickness (Davis, 1946; Jones and others, 1948) suggest that there has been widespread subsidence in the water-conservation areas over the past 50 years. The northern parts of individual water-conservation areas may still experience some minor subsidence. The southern or downstream parts of the impoundments are generally wetter and may be accumulating peat (Craft and Richardson, 1993a, 1993b), very gradually increasing in elevation. In the drained agricultural and urban areas, subsidence is an ongoing process, except where the peat has already disappeared entirely.

SUBSIDENCE CLOUDS THE FUTURE OF AGRICULTURE

The Everglades agricultural area is now mainly devoted to sugarcane, with considerably smaller areas used for vegetables, sodgrass, and rice. The value of all agricultural crops is currently about $750 million (Snyder and Davidson, 1994).

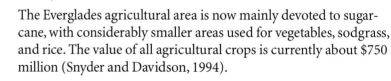

The eventual demise of agriculture in the Everglades has been predicted for some time (Douglas, 1947; Stephens and Johnson, 1951). The agriculture depends upon a relatively thin, continually shrinking layer of peat soil that directly overlies limestone bedrock. Agronomists have known for many decades that peat-rich soils (histosols), which form in undrained or poorly drained areas, will subside when drained and cultivated. The causes include mechanical compaction, burning, shrinkage due to dehydration, and most importantly, oxidation of organic matter. Oxidation is a microbially mediated process that converts organic carbon in the soil to (mainly) carbon dioxide gas and water.

Through photosynthesis, vegetation converts carbon dioxide and water into carbohydrates. Under natural conditions, aerobic microorganisms converted dead plant material (mostly sawgrass root) to peat during brief periods of moderate drainage. Vegetative debris was deposited faster than it could fully decompose, causing a gradual increase in peat thickness. In what is now the Everglades agricultural area, a delicate balance of 9 to 12 months flood and 0 to 3 months slight (0 to 12 inches) drainage for about 5,000 years, with sawgrass the dominant species, led to a peat accretion rate of about 0.03 inches per year. Drainage disrupted this balance so that, instead of accretion, there has been subsidence at a rate of about 1 inch per year.

Peat soils may virtually disappear

Rates of subsidence in the Everglades are slower than those in the Sacramento-San Joaquin Delta of California, the other major area of peat-oxidation subsidence in the United States; in the Delta, average subsidence rates have been up to 3 inches per year. However, the preagricultural peat thickness was much greater in the Delta (up to 60 feet) than in the Everglades, where initial thicknesses were less than 12 feet. The subsidence rates observed in the Everglades are similar to those observed in the deep peat soils of the English fens during the past 100 years (Lucas, 1982; Stephens and others, 1984).

In the Everglades agricultural area, the initial peat thickness tapered southward from approximately 12 feet near Lake Okeechobee to about 5 feet near the southern boundary. In 1951, Stephens and

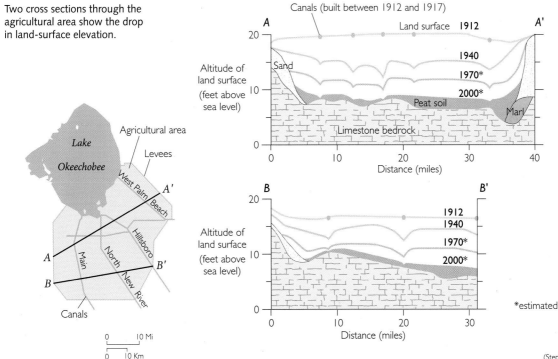

Two cross sections through the agricultural area show the drop in land-surface elevation.

(Stephens and Johnson, 1951)

A ditch excavation east of Belle Glade shows peat soil overlying limestone bedrock.

Johnson extrapolated contemporary subsidence trends to predict that by the year 2000 the peat soil would be less than 1 foot thick in about half of the area. They further inferred that much of the area will by then have gone out of agricultural production, assuming that cultivation would not be possible with less than 1 foot of soil over limestone bedrock.

Although the extrapolation of peat thickness done by Stephens and Johnson (1951) appears consistent with measurements made in 1969 (Johnson, 1974), 1978 (Shih and others, 1979), and 1988 (Smith, 1990), little land has yet been retired from sugarcane. One reason is that farmers have managed to successfully produce cane from only 6 inches of peat, by first piling it in windrows to allow successful germination. It also appears possible that the rate of subsidence has slowed somewhat (Shih and others, 1997), due to the combined effects of an increasing nonorganic (mineral) content in the remaining soil, a thinner unsaturated zone dictated by the decreasing soil depth and, perhaps, an increasing abundance of more recalcitrant forms of organic carbon.

The soil-depth predictions of Stephens and Johnson (1951) may prove to have been somewhat pessimistic, but it is clear that agriculture as currently practiced in the Everglades has a finite life expectancy, likely on the order of decades. Extending that life expectancy would require development of an agriculture based on water-tolerant crops that accumulate rather than lose peat (Porter and others, 1991; Glaz, 1995).

SUBSIDENCE COMPLICATES ECOSYSTEM RESTORATION

In a wetland area where natural hydraulic gradients were on the order of inches per mile, and one half-foot land-surface altitude differences are ecologically significant, the fact of several feet of land subsidence substantially complicates ecosystem-restoration efforts.

Subsidence makes true restoration of the Everglades agricultural area itself technically impossible, even in the event that it were po-

Canals and a levee separate constructed wetland from the agricultural area to the right.

litically and economically feasible. Land there that once had a mean elevation less than 20 feet above sea level has been reduced in elevation by an average of about 5 feet. Differential subsidence has significantly altered the slope of the land, precluding restoration of the natural, shallow sheet-flow patterns. If artificial water management and conveyance were now to cease, nature would likely reclaim the land as a lake, rather than the predevelopment sawgrass plains. With removal of the "sponge" of peat and native vegetation, the agricultural area has also lost most of its ability to naturally filter, dampen, and retard storm flows. Other strong impediments to restoration of the Everglades agricultural area include loss of the native seed bank, accumulations of agricultural chemicals in the soil, and the potential for invasion by aggressive exotic species.

Subsidence will also complicate efforts to manage the water-conservation areas to the east and south in a more natural condition. For example, the wetlands in these areas are speckled with tree islands, which are an important ecosystem component. Though definitive data are lacking, these tree islands likely have subsided, possibly more than the surrounding area. Thus, restoration of the water-conservation areas will require careful management of water levels in a depth range sufficient to promote appropriate wetland species without further damaging tree islands.

CAREFUL WATER MANAGEMENT IS A KEY TO THE FUTURE

Because of peat loss, agriculture as currently practiced in the Everglades will gradually diminish over the next decades. R.V. Allison, the first head of the Everglades Research Station, likened the peat soil to "the cake which we cannot eat and keep at the same time." His confident prediction that

"As the use of Everglades lands for agricultural purposes approaches the sunset of ... production, there is little doubt that transition into a wildlife area of world fame will follow, perhaps in an easy and natural manner."   —Allison, 1956

now seems overly optimistic. This is still a possible scenario but, as we have noted, the result would be very different from the natural system, due to subsidence. There are also alternative possibilities, including urban development or invention of a sustainable agriculture.

A sustainable agriculture in the Everglades would require at least zero subsidence and, optimally, some peat accretion. Glaz (1995) discussed a program of genetic, agronomic, and hydrologic research aimed at gradually (over a period of 20 to 40 years) making a currently used sugarcane-rice rotation sustainable. Achievement of this goal may prove difficult. However, documented water tolerance of sugarcane (Gascho and Shih, 1979; Kang and others, 1986; Deren and others, 1991), a recently discovered explanation for this water tolerance (Ray and others, 1996), and rapid gains in molecular genetics combine to suggest that substantial reductions in subsidence might be attainable.

A tree island in the Everglades

Even in the complete absence of agriculture in the Everglades, the existing pattern of urban development and land subsidence would prevent restoration of the natural flow system. Engineered water management and conveyance will be required indefinitely. Land subsidence over a large area south of Lake Okeechobee has created a significant trough within the natural north-south flow system, thereby preventing restoration of natural sheet-flow and vegetation patterns.

The Everglades are currently the subject of a major Federal-State ecosystem restoration effort. "Restoration" is perhaps a misnomer, as the focus of this effort is on more natural management of the remaining 50 percent of the Everglades wetlands, not on regaining the 50 percent that has been converted to urban and agricultural use. Even improving the natural functioning of the remaining wetlands will be a complex problem, due to the lost spatial extent, the hydrologic separation from Lake Okeechobee, and land subsidence. The Everglades will likely continue to be an intensively managed system. However, much as the major engineering effort in the 1950s and 1960s halted the destructive fires and saltwater intrusion of preceding decades, the current restoration effort has the potential to halt and reverse more recent environmental degradation. A major challenge will be to deliver water from Lake Okeechobee through the extensive subsided areas so that it arrives in the undeveloped southern Everglades at similar times, in similar quantities, and with similar quality, as it did prior to drainage and subsidence.

PART III

Collapsing Cavities

The Retsof Salt Mine Collapse

Sinkholes, West-Central Florida

S udden and unexpected collapse of the land surface into subsurface cavities is arguably the most hazardous type of subsidence. Such catastrophic subsidence is most commonly triggered by ground-water-level declines caused by pumping, or by diversion of surface runoff or ground-water flow through susceptible rocks. Though the collapse features tend to be highly localized, they can introduce contaminants to the aquifer system and, thereby, have lasting regional impacts. Collapse features tend to be associated with specific rock types having hydrogeologic properties that render them susceptible to the formation of cavities. Human activities can facilitate the formation of subsurface cavities in these susceptible materials and trigger their collapse, as well as the collapse of preexisting subsurface cavities.

In terms of land area affected, underground mining accounts for about 20 percent of the total land subsidence in the United States, and most of this fraction is associated with underground mining for coal. Subsidence over underground coal workings develops as a gradual downwarping of the overburden into mine voids and is generally unrelated to subsurface water conditions. Underground salt and gypsum mines are also subject to downwarping of the overburden, but these evaporite minerals are also susceptible to rapid and extensive dissolution by water. Salt and gypsum are, respectively, almost 7,500 and 150 times more soluble than limestone, the rocktype often associated with catastrophic sinkhole formation and the distinctively weathered landscapes collectively known as karst. Here, we consider only the collapse of cavities that form in soluble rocks such as salt, gypsum, and limestone.

Formation of subsurface cavities by dissolution requires: 1) bedrock composed in large part of soluble minerals; 2) a water source that is unsaturated with respect to these minerals and, therefore, can dissolve them; 3) an energy source in the form of a hydraulic gradient to move the water through the rock; and 4) an outlet for the escaping, mineralized water. Once a through-flowing passage develops in the soluble rock, erosion and further dissolution enlarges the pas-

This sinkhole in Kansas was formed by collapsed evaporite rocks.

In western Kansas dissolution of gypsum and salt beds several hundred feet below the surface caused the sudden formation of the Meade Sink in March 1879. The hole was about 60 feet deep and 610 feet in diameter and filled with saltwater. Today the sink has partly filled with sediment and is usually dry.

(Kansas Geological Survey)

sage, further enhancing the throughflow. Once established, subsurface cavities may provide habitat for populations of species specially adapted to cave environments—a cave ecosystem. The interaction between these biological communities and the mineral substrate of the host cavities may further enhance mineral dissolution and cavity enlargement through the production of acid metabolites.

EVAPORITE ROCKS CAN FORM CAVITIES WITHIN DAYS

Evaporites are sediments deposited from natural waters that have been concentrated as a result of evaporation. Evaporite rocks such as salt and gypsum underlie about 35 to 40 percent of the contiguous United States. Natural solution-related subsidence has occurred in each of the major salt basins (Ege, 1984), perhaps most notably in the Permian basin of Texas, New Mexico, Oklahoma, and Kansas and the smaller Holbrook basin of northeast Arizona. Although evaporites underlie most of the Michigan-Appalachian and Gulf Coast basins, naturally forming collapse features are much less common in these areas. Human-induced collapse cavities are relatively uncommon in gypsum deposits, and more likely to develop above salt deposits, where they are associated with both purposeful and accidental dissolution of salt.

Salt and gypsum underlie about 40 percent of the contiguous United States.

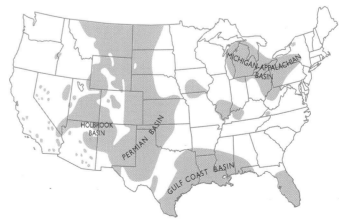

(Martinez and others, 1998)

THE RETSOF SALT MINE COLLAPSE

Widespread subsidence occurred after a mine collapse in the Genesee Valley, New York

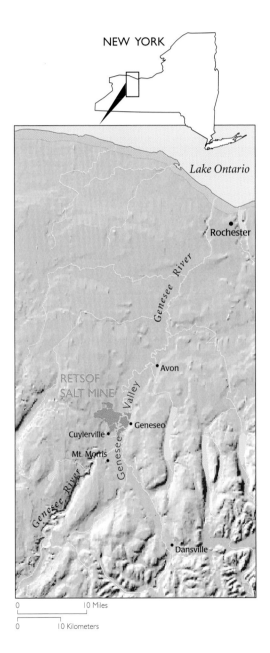

NEW YORK

Lake Ontario

Rochester

Genesee River

Avon

RETSOF SALT MINE

Genesee Valley

Cuylerville

Geneseo

Mt. Morris

Genesee River

Dansville

0   10 Miles

0   10 Kilometers

On March 12, 1994, at 5:43 a.m. (local time), an apparent earthquake of magnitude 3.6 centered near Cuylerville, New York, woke residents and registered on seismographs 300 miles away. Prompted by a call placed from a local resident, the USGS National Earthquake Information Center confirmed that a seismic event had occurred near Cuylerville and immediately notified State emergency services offices in New York who, in turn, notified the Livingston County Sheriff's Department. The Sheriff's Department contacted the Retsof Mine, which, except for some limited subsurface maintenance activity, had suspended active mining that weekend.

Mine officials discovered that a 500- by 500-foot section of shale roof rock some 1,200 feet below land surface had collapsed in a part of the mine known as room 2-Yard South. Mine officials detected methane and hydrogen sulfide gases, and ground water was flowing into the mine from the roof collapse area at nearly 5,000 gallons per minute.

This collapse began a series of events that would eventually lead to the further collapse and complete flooding of the mine, large declines in local ground-water levels, degradation of potable ground-water supplies, land subsidence, release of natural gases (methane and hydrogen sulfide) to the atmosphere, and other detrimental effects on the cultural resources and infrastructure in this part of the Genesee Valley.

William M. Kappel, Richard M. Yager, and Todd S. Miller
U.S. Geological Survey, Ithaca, New York

Seismogram recorded at Cuylerville, New York March 12, 1994

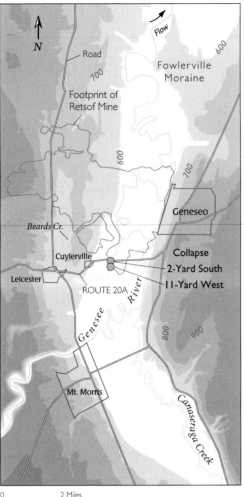

0   2 Miles

0   2 Kilometers

Land-surface altitude
(feet above sea level)

SALT MINING HAS A LONG HISTORY IN THE GENESEE VALLEY

Salt mining (both salt-solution and rock-salt mining) began in the Genesee Valley in the early 1880s, and in 1884 the Empire Salt Company excavated a shaft to extract rock salt from seams 900 feet below land surface. In 1885 the Empire Salt Company was renamed the Retsof Mine Company and the Village of Retsof was founded near the mine shaft. During the next 110 years, the mine grew to become the largest salt-producing mine in the United States and the second largest in the world. Before the initial collapse in March 1994, the mine encompassed an underground area of more than 6,000 acres, and the mine footprint (outer edge of mined area) extended over an area of nearly 10 square miles.

At the time of the collapse, the Retsof Mine was owned by Akzo-Nobel Salt Incorporated (ANSI), and, during the winter of 1993–94, operated at full capacity to meet demands for road salt throughout the northeastern United States. Prior to its closure, the Retsof Mine played a major role in the Livingston County economy, providing more than 325 jobs with an annual payroll in excess of $11 million and estimated annual gross sales of more than $70 million (NYSDEC, 1997). During the 17 months following the collapse, mining operations shifted to the northern, high end of the mine in a race to salvage mineable salt before the mine flooded. The Retsof Mine ceased operations on September 2, 1995, and by December, 21 months after the initial collapse, the mine was completely flooded.

THE COLLAPSE TRIGGERED A SERIES OF LOCAL EVENTS

Four months before the collapse, in November 1993, room 2-Yard South was abandoned because of concerns over large and increasing rates of "convergence" or reduction of the opening between the floor and ceiling of the room. (A new mining technique, "yielding pillar," was used in this area in response to floor buckling and roof collapse, which was occurring with greater frequency in the south-

This cross-sectional schematic shows how water from the basal aquifer entered the mine through the collapsed area. After 21 months the salt mine was completely flooded.

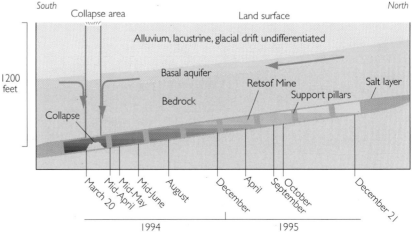

This northwest aerial view shows sinkholes above room 11-Yard West (left foreground) and room 2-Yard South (right center, partially obscured by trees). Circle indicates location of the Retsof main plant area, some 4 miles northwest of the sinkholes.

(Ron Pretzer, LUXE, May 1994)

A sinkhole developed above room 11-Yard West and filled with water from Beards Creek.

(Richard Young, Geological Sciences, SUNY Geneseo, June 1994)

The roadbed of Route 20A was fractured on the east side of the collapsed bridge over Beards Creek. This view is above the room 2-Yard South collapse area looking west, toward the former Hamilton farm house (subsequently purchased by ANSI).

(Richard Young, Geological Sciences, SUNY Geneseo, April 12, 1994)

ern end of the mine.) After the March 12, 1994, collapse of room 2-Yard South, ground water flowed into the previously dry mine at a rate of about 7 million gallons per day, dissolving residual rock salt and filling the lowest, downdip levels of the mine with saturated brine. ANSI monitored the concentration of hazardous gases and the encroaching water level in the mine as the shoreline in the mine steadily moved northward.

Local governmental officials had posted warning signs at the Route 20A bridge over Beards Creek the day before the collapse because of small bumps in the pavement on the bridge approach sections. There is some anecdotal evidence that, several days earlier, local travelers had noticed a change in the smoothness of the roadbed near the bridge, suggesting a surface expression of the underground convergence that led to abandonment of room 2-Yard South several months earlier.

Within days of the collapse, impacts on the glacial and bedrock aquifer systems and on the land surface were reported on an expanding scale. Some homes had sustained structural damage due to the initial earth tremors and, within 1 week of the collapse, residents along Wheelock Road, south of Cuylerville and southwest of the mine, reported that several water wells had gone dry (NYSDEC, 1997). The USGS and the Livingston County Health Department began monitoring ground-water levels and streamflow in the area.

On April 6, a 200-foot diameter by 20-foot deep, cone-shaped sinkhole appeared along the channel of Beards Creek, immediately above the room 2-Yard South collapse zone, just south of the Route 20A bridge. And 2 weeks later, accompanied by additional earth tremors, this sinkhole expanded to about 600 feet in diameter.

On April 8, seismic events indicated a roof collapse in mine room 11-Yard West, south of and adjacent to room 2-Yard South. Following this collapse, ground-water inflow to the mine increased to about 22 million gallons per day. An expanding sinkhole developed over 11-Yard West on May 25, 1994. The sinkhole was about 50 feet

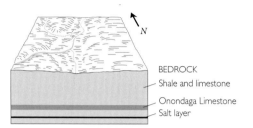

The ancient Genesee River crossed sedimentary rocks.

BEDROCK
Shale and limestone
Onondaga Limestone
Salt layer

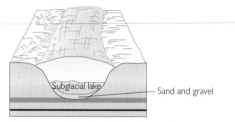

Sand and gravel

Glaciers scoured the bedrock, leaving a wide, deep valley that did not always follow the course of the Genesee River. At times the glaciers covered the entire area. During periods of glacial retreat, subglacial lakes formed and sediment was deposited.

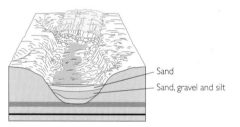

Sand
Sand, gravel and silt

The periodic retreat and advance of glaciers left behind mounds of debris (moraines) and thick glacial deposits (drift).

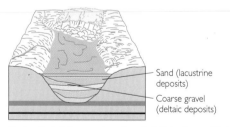

Sand (lacustrine deposits)
Coarse gravel (deltaic deposits)

During deglaciation a series of proglacial lakes formed that deposited lake (lacustrine) sediments on the valley floor.

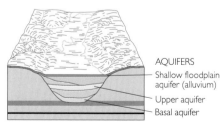

AQUIFERS
Shallow floodplain aquifer (alluvium)
Upper aquifer
Basal aquifer

After glaciation alluvial (floodplain) gravel, sand, and silt were deposited on top of the glacial sediments.

deep, about 200 feet in diameter, and immediately filled with water captured from Beards Creek. Over time this sinkhole grew to about 800 feet in diameter.

THE NATURAL HISTORY OF THE GENESEE VALLEY SET THE STAGE FOR WIDESPREAD DAMAGE AFTER THE COLLAPSE

Current knowledge of the occurrence and flow of ground water and the complex stratigraphy of the glacial aquifer system in Genesee Valley is sparse. Prior to the collapse, the hydrogeologic framework of the valley-fill materials had not been investigated in detail. Since the mine collapse, several studies have addressed the hydrogeologic framework (Nittany Geoscience, 1995; Alpha Geoscience, 1996), but insufficient data exist to thoroughly characterize the interconnections among glacial units and bedrock aquifer zones.

The Genesee Valley from Dansville to Avon, New York, includes the Canaseraga Creek Valley and, from Mt. Morris northward, the Genesee River Valley. The valley formed as a result of several geologic processes including the ancestral uplift and stream erosion of gently dipping Paleozoic sedimentary rocks, followed by periods of glaciation in which ice scoured and modified the bedrock topography, leaving behind unconsolidated sediments. Recently, stream erosion and deposition added about 50 feet of alluvium (gravel, sand, and silt) to the glacial sediments.

The unconsolidated glacial sediments that fill the Genesee Valley were deposited during cycles of glacial advances and retreats. Glaciers several thousands of feet thick deepened and widened the valley. About 12,000 years ago the most recent glacier retreated from the valley, leaving behind thick glacial deposits. Where the glaciers paused and the ice melted, mounds of glacial debris, called end moraines, were deposited at the frontal (southern) ice margin. The melting ice produced large volumes of water that transported, sorted, and deposited boulders, gravel, cobbles, sand, silt, and clay and carried these sediments in meltwater streams to the south. Proglacial lakes existed in the glacially-deepened valley between the valley walls and the receding glacier during most of the glacial period.

During deglaciation, outlets low enough to drain the proglacial lakes did not exist until the ice margin was 10 to 12 miles north of Geneseo. During this period, the present Genesee River and Canaseraga Creek watersheds drained to the north, toward the glacier, into a series of progressively lower proglacial lakes. The final and lowest proglacial lake formed when the ice deposited the Fowlerville moraine, which extends from about 4.5- to 8-miles north of the collapse area. Water ponded in the Genesee Valley south of the Fowlerville moraine, depositing lake sediments on the valley floor. Eventually the lake drained as the Genesee River cut a channel in the Fowlerville moraine (Young, 1975). As much as 700 feet of glacially derived gravel, sand, silt, and clay were deposited in a subglacial and glaciolacustrine (glacial lake) environment.

The buried bedrock surface follows the slope of the resistant sedimentary carbonate beds of the Onondaga Limestone, dipping approximately 42 feet per mile to the south. Overlying the bedrock surface is a thickening wedge of glacial valley-fill sediments that ranges from a few hundred feet thick on the north, near the Fowlerville Moraine, to about 750 feet thick in the deepest part of the valley near Sonyea. South of Sonyea, the valley fill thins. Ground water in the glacial deposits and portions of the underlying carbonate bedrock has been the primary source for the inflows to the flooded Retsof Mine. The fine-grained lake silt and clay closer to the land surface form a barrier between the alluvial and deeper glacial aquifers.

Water-bearing zones are found within the fractures and bedding planes near the top of the Onondaga Limestone at the base of the valley fill. Another water-bearing zone is found at the contact between the Onondaga Limestone and the underlying Bertie Limestone. Few valley wells tap bedrock, and the most productive wells completed in the Onondaga and Bertie Limestones seldom produce more than several tens of gallons per minute (Dunn, 1992). The Bertie Limestone subcrops beneath the valley floor north of the Fowlerville moraine, under several hundred feet of glacial sediment, and is generally considered a divide between fresher water above and a more mineralized water below.

The principal aquifer in the valley appears to occur at the base of the valley fill. The relatively thin basal aquifer is composed of sand and gravel deposited on top of the Onondaga Limestone in the central and northern parts of the valley and on top of the low-permeability Devonian shales to the south. The hydraulic connection between the basal aquifer and the underlying bedrock units throughout the valley is poorly understood, but the connection is generally assumed to be better in the northern half of the valley, where the aquifer is in direct contact with the weathered and fractured top of the Onondaga Limestone. Under natural conditions ground water flows upward from the Onondaga to the basal aquifer. Though the basal aquifer is generally overlain by lower-permeability

This view shows the Upper Genesee Valley looking east.

(Richard Young, Geological Sciences, SUNY Geneseo)

Water from the basal aquifer entered the mine through the collapsed area.

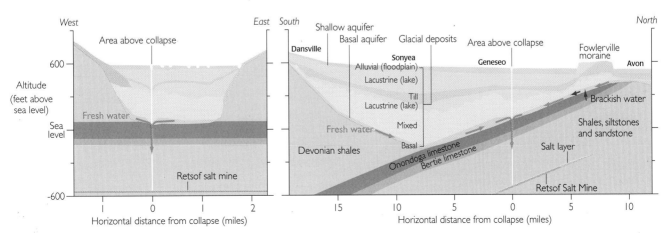

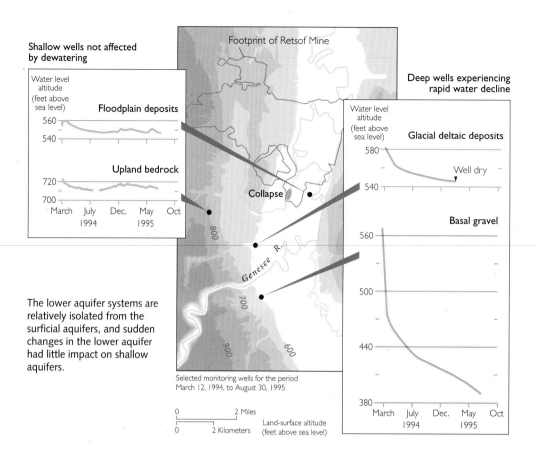

Shallow wells not affected by dewatering

Water level altitude (feet above sea level)

Floodplain deposits

Upland bedrock

The lower aquifer systems are relatively isolated from the surficial aquifers, and sudden changes in the lower aquifer had little impact on shallow aquifers.

Deep wells experiencing rapid water decline

Water level altitude (feet above sea level)

Glacial deltaic deposits

Well dry

Basal gravel

Footprint of Retsof Mine

Collapse

Genesee R.

Selected monitoring wells for the period March 12, 1994, to August 30, 1995

Land-surface altitude (feet above sea level)

This view of the Genesee Valley floodplain was taken above the southern end of the mine looking southwest.

(Richard Young, Geological Sciences, SUNY Geneseo, 1995)

glacial drift, in some areas north of the mine more permeable layers have been reported within the glacial deposits.

Some wells in the valley are completed within the glacial deposits, and some wells completed in the deeper basal aquifer are also screened in the glacial deposits, an indication that there is locally enhanced permeability at intermediate depths. There appears to be a vertical hydraulic connection between the basal aquifer and the permeable zones in the glacial deposits, based upon recent data from ground-water monitoring wells, but the areal extent of these vertically connected zones is unknown.

Shallow ground water occurs in the alluvial deposits found to a depth of 50 feet below the valley floor. The water table in the alluvium is generally less than 15 feet below land surface, and is in hydraulic connection with the Genesee River, Canaseraga Creek, and other tributaries on the valley floor. Other shallow ground water occurs in the Fowlerville Moraine deposits. Most recharge and discharge of the Genesee Valley aquifer system occurs between the Genesee River, its tributaries, and the shallow water-table aquifer in the alluvium (Nittany Geoscience, 1995). Water levels in wells completed in the alluvium were not affected by the mine collapse.

After the mine collapse, most of the inflows to the mine probably came from storage in the basal aquifer and the glacial deposits through the collapse areas above rooms 2-Yard South and 11-Yard

West. Water levels in wells began declining almost immediately near the collapse zones. Water levels continued to decline rapidly through 1994, and more slowly in 1995, until the mine was completely flooded in January 1996. By then, water levels had fallen more than 350 feet in some wells near the collapse zones. In total, an estimated 42,000 acre-feet of ground water invaded the mine.

The basal aquifer is relatively isolated from surficial sources of recharge and discharge, and changes in the lower part of the aquifer system are not likely to have immediate or significant impact on the shallow sources. However, the rate of ground-water drainage into the mine far exceeded the estimated rate of recharge to the deeper subsurface aquifers, and it is expected that it will take a decade or longer for ground-water levels to recover throughout the aquifer system.

IMPACTS OF THE COLLAPSE WERE OBSERVED MILES AWAY

The effects of the collapse include, but are not limited, to the following:

- Reduced air quality and public-safety issues resulting from the emanation of methane and hydrogen-sulfide gases

- The loss of potable water supplies—both a reduction of quantity and degradation in quality and

- Short- and long-term land subsidence

Natural gas was vented into the environment

Soon after the mine began to flood and water levels in the basal aquifer were lowered, natural gas in the form of hydrogen sulfide (odor of rotten eggs) and methane (odorless, combustible) began exsolving from ground water—just as carbon dioxide comes out of solution after a bottle of soda is opened. In the area of the collapse, lowered water levels allowed natural gas to escape through test wells drilled near the collapse area and preexisting domestic water-supply wells several miles farther to the southeast. In September 1994 the State Department of Environmental Conservation ordered ANSI to develop a natural-gas monitoring and response plan. By May 1995 the County and State Health Departments required ANSI to flare-off (burn) gas from several collapse-area wells to reduce the odor and protect the health and safety of residents living in Cuylerville and the surrounding area.

Potable water supplies were diminished

Although some shallow alluvial wells near the mine were unaffected, some domestic wells along the margins of the valley and in the deeper zones of the Genesee Valley aquifer system experienced lowered water levels, and some wells went dry. The rate of water-level decline varied: water levels declined 20 feet or more along Wheelock Road (about 1 mile southwest of the mine) within days of the collapse, whereas water levels gradually declined 50 feet or more in the

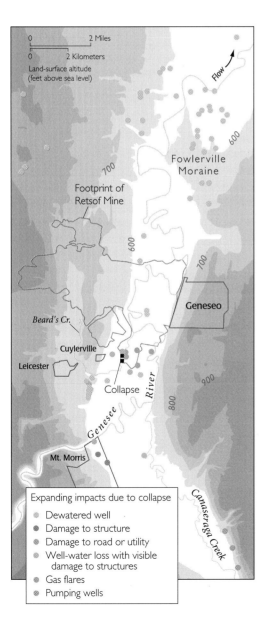

0   2 Miles
0   2 Kilometers
Land-surface altitude
(feet above sea level)

Flow

Fowlerville
Moraine

Footprint of
Retsof Mine

Geneseo

Beard's Cr.

Cuylerville

Leicester

Collapse

Genesee River

Mt. Morris

Canaseraga Creek

Expanding impacts due to collapse
- Dewatered well
- Damage to structure
- Damage to road or utility
- Well-water loss with visible damage to structures
- Gas flares
- Pumping wells

Local subsidence occurred due to dissolution of salt pillars by freshwater inflow March 1994 to March 1996.

More extensive subsidence due to closure of the salt cavity is projected.

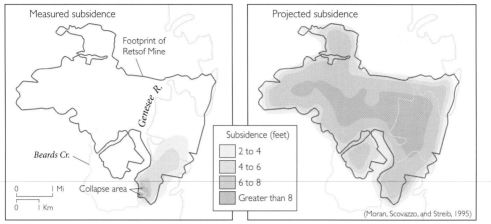

A well flares methane and hydrogen sulfide gases from a fracture zone on the eastern margin of the room 2-Yard South sinkhole.

(Richard Young, Geological Sciences, SUNY Geneseo, 1995)

Fowlerville area (about 6 miles north of the mine) and in Mt. Morris (about 4 miles south of the mine) for 2 years following the collapse. Pursuant to an agreement between ANSI, Livingston County, and the State of New York, ANSI has been supplying water to residents whose wells have gone dry and where water quality has deteriorated.

The effects of ground-water flow to the mine extend more than 10 miles north and south of the collapse area. Following the mine collapse and lowering of ground-water levels, highly mineralized ground water has apparently migrated into freshwater supplies. There are two potential sources: a deep-basin brine that migrates upward along bedding-plane fractures within the Bertie Limestone to the intersection of the Bertie outcrop and the basal aquifer, and a halite (rock salt) component, which may be introduced through older natural-gas or salt-solution wells within the Fowlerville moraine. The mineralized ground water flows downdip (to the south) through the basal aquifer toward the mine collapse area, an apparent reversal of the pre-collapse hydraulic gradient. Presently, salinity is increasing in Fowlerville Moraine wells, south of where the Bertie outcrops and is in contact with the basal aquifer.

Several types of subsidence were observed

Subsidence damage related to the mine collapse includes:

• The creation of 2 large sinkholes

• The temporary loss of State Route 20A through Cuylerville

• Structural damage to homes and businesses and

• Damage to agricultural lands, public utilities, and cultural resources.

Besides the catastrophic formation of sinkholes over rooms 2-Yard South and 11-Yard West, the damage involves three other types of subsidence, which are important at different scales.

An earth fissure ruptures a field above the western edge of the mine, northwest of 2-Yard South sinkhole. The field surrounding the fissure subsided almost 1 foot.

(Richard Young, Geological Sciences, SUNY-Geneseo, March 1995)

The lowering of water levels in areas beyond the mine footprint may eventually cause aquifer-system compaction and land subsidence.

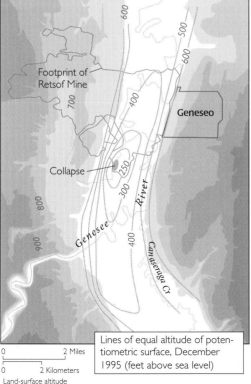

Footprint of Retsof Mine

Geneseo

Collapse

Genesee River

Canaseraga Cr

Lines of equal altitude of potentiometric surface, December 1995 (feet above sea level)

0 — 2 Miles
0 — 2 Kilometers
Land-surface altitude (feet above sea level)

The first type of subsidence that normally occurs over any mined-out area is due to the slow closure of the mine opening. Mining engineers expect the land overlying the Retsof Mine footprint to subside about 8 to 9 feet over the next 100 to 200 years (Van Sambeek, 1994). Most of the estimated subsidence is expected to be realized during the next 100 years (Shannon and Wilson, 1997). Differential subsidence is expected along the margins of the mine, where adjacent areas will subside nonuniformly. This creates stresses within the land mass, which may rupture the surface or subsurface. Some horizontal movement of land surface in these areas is expected, as well as some tilting of the land surface toward the mine. Structures located in these regions may continue to be prone to damage as the mine subsidence evolves.

A second type of subsidence seen near the collapse area and farther to the north and east was caused by the flow of ground water into the mine and resultant dissolution of unmined salt. Fresh ground water, less dense than saltwater, entered the mine cavity quickly, and preferentially dissolved the salt along the mine roof. As the mine roof collapsed, it allowed the freshwater to dissolve more salt in the supporting salt pillars and, over time, left large areas without roof support. This type of subsidence evolved rapidly as many salt pillars were quickly dissolved by the large inflow of freshwater, and subsidence in this area was greater and occurred sooner than would be expected for a dry mine situation (Van Sambeek, 1996). When the mine filled with saturated brine, this type of subsidence ceased.

The third type of subsidence to occur in the Genesee Valley is due to the dramatic lowering and anticipated slow recovery of ground-water levels in the confined-aquifer system. This type of subsidence is due to aquifer-system compaction that typically accompanies the depletion of alluvial aquifer systems. The ground-water level declines experienced after the mine-roof collapse—more than 350 feet near the collapse and as much as 50 feet as far as 8 miles away—is sufficient to cause measurable elastic compression of the glacial sediments of the Genesee Valley aquifer system. It is possible that the large stresses imposed on the aquifer-system skeleton by the large drawdowns may have caused some inelastic, and largely irreversible, compaction of aquitards in the Genesee Valley, but currently this effect is presumed to be small. Aquifer-system compaction may contribute to land subsidence on a spatial scale larger than the mine footprint, especially in regions where large changes in ground-water levels persist and where the glacial deposits contain an appreciable thickness of fine-grained, more compressible sediments. It is possible that the valley floor may continue to be affected by residual compaction long after ground-water levels have fully recovered in the aquifers (Riley, 1969). An accurate evaluation of the magnitude, timing, and areal extent of land subsidence due to aquifer-system compaction will depend on more detailed knowledge of the hydrogeology of the Genesee Valley.

CONTINUING STUDIES WILL ASSESS FUTURE IMPACTS

The long-term lowering of aquifer hydraulic heads creates the potential for permanent compaction of the aquifer system and additional land subsidence. The distribution of compressible sediments and their mechanical behavior need to be better understood in order to predict potential impacts. The sources of poor-quality water and potential paths of migration in the aquifers also need to be assessed in order to evaluate and predict changes in ground-water quality throughout the Genesee Valley.

The USGS is currently implementing conceptual and numerical models of ground-water flow in Genesee Valley to assist in determining the impact of mine flooding on the regional aquifer system. Drainage of ground water into the collapse areas is being simulated using data collected by ANSI consultants; the State Departments of Law, Environmental Conservation, and Health; Livingston County; local citizens; the USGS; and others. The models will provide insight into the problems of lowered ground-water levels, land subsidence caused by aquifer-system compaction, and migration of mineralized ground water.

SINKHOLES, WEST-CENTRAL FLORIDA

A link between surface water and ground water

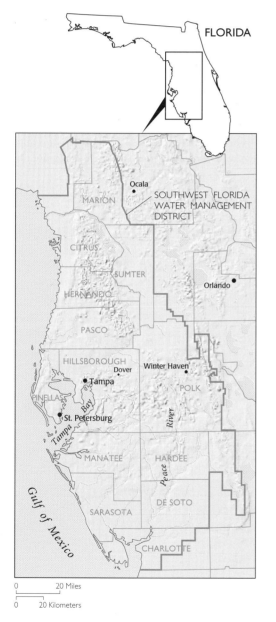

0   20 Miles

0   20 Kilometers

Sinkholes are a common, naturally occurring geologic feature and one of the predominant landforms in Florida, where they pose hazards to property and the environment. Although many new sinkholes develop naturally, in west-central Florida and elsewhere, their increasing frequency corresponds to the accelerated development of ground-water and land resources. Usually little more than a nuisance, new sinkholes can sometimes cause substantial property damage and structural problems for buildings and roads. Sinkholes also threaten water and environmental resources by draining streams, lakes, and wetlands, and creating pathways for transmitting surface waters directly into underlying aquifers. Where these pathways are developed, movement of surface contaminants into the underlying aquifer systems can persistently degrade ground-water resources. In some areas, sinkholes are used as storm drains, and because they are a direct link with the underlying aquifer systems it is important that their drainage areas be kept free of contaminants. Conversely, when sinkholes become plugged, they can cause flooding by capturing surface-water flow and can create new wetlands, ponds, and lakes.

Most of Florida is prone to sinkhole formation because it is underlain by thick carbonate deposits that are susceptible to dissolution by circulating ground water. Florida's principal source of freshwater, ground water, moves into and out of storage in the carbonate aquifers—some of the most productive in the nation. Development of these ground-water resources for municipal, industrial and agricultural water supplies creates regional ground-water-level declines that play a role in accelerating sinkhole formation, thereby increasing susceptibility of the aquifers to contamination from surface-water drainage. Such interactions between surface-water and ground-water resources in Florida play a critical and complex role in the long-term management of water resources and ecosystems of Florida's wetlands (see Florida Everglades in Part II of this Circular).

Ann B. Tihansky
U.S. Geological Survey, Tampa, Florida

Reported sinkholes
from 1960 to 1991

(Wilson and Shock, 1996)

SINKHOLES ARE A NATURALLY OCCURRING FEATURE IN THE FLORIDA LANDSCAPE

The exposed land mass that constitutes the Florida peninsula is only part of a larger, mostly submerged carbonate platform that is partially capped with a sequence of relatively insoluble sand and clay deposits. Siliciclastic sediments (sand and clay) were deposited atop the irregular carbonate surface, creating a blanket of unconsolidated, relatively insoluble material that varies in composition and thickness throughout the State. In west-central Florida, the relation between the carbonate surface and the mantling deposits plays an important role in the circulation and chemical quality of ground water and the development of landforms. Sinkhole development depends on limestone dissolution, water movement, and other environmental conditions. Limestone dissolution rates (on the order of millimeters per thousand years) are highest in areas where precipitation rates are high. Cavities develop in limestone over geologic time and result from chemical and mechanical erosion of material (Ford and Williams, 1989).

Dissolving carbonate rocks create sinkholes and other features

The soluble limestones and dolomites that constitute the carbonate rocks are sculpted by dissolution and weathering processes into a

There appears to be an increasing frequency of sinkholes, although the statistics may be affected by reporting biases.

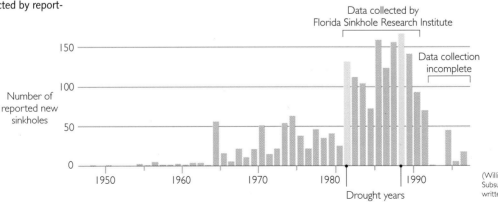

Data collected by
Florida Sinkhole Research Institute

Data collection
incomplete

Number of
reported new
sinkholes

150

100

50

0

1950   1960   1970   1980   1990

Drought years

(William L. Wilson,
Subsurface Evaluations, Inc.,
written communication, 1997)

distinct geomorphology known as karst. Features characteristic of karst terranes are directly related to limestone dissolution and ground-water flow and include sinkholes, springs, caves, disappearing streams, internally drained basins, and subsurface drainage networks. Dissolution cavities can range in size from tiny vugs to gigantic caverns. As these enlarging voids coalesce and become hydraulically interconnected, they greatly enhance the movement of ground water, which can perpetuate further dissolution and erosion.

On a local scale, the caverns and cave networks can form extensive conduit systems that convey significant ground-water flow at very high velocities (Atkinson, 1977; Quinlan and others, 1993). On a regional scale, the many interconnected local-scale features can create a vast system of highly transmissive aquifers that constitute a highly productive ground-water resource.

Changes in sea level helped develop karst terranes

Karst is well-developed in the carbonate rocks throughout the Florida carbonate platform. Throughout recent geologic time, fluctuations in sea level have alternately flooded and exposed the platform, weathering and dissolving the carbonate rocks. During the Ice Ages, an increased proportion of the Earth's water was frozen in polar ice and continental glaciers, lowering sea level along the Florida peninsula by 280 to 330 feet as recently as 18,000 years ago. The sea-level low stands exposed the great carbonate platforms of the Gulf of Mexico and the Caribbean Sea to karst processes. The lower sea-level stands were accompanied by lower ground-water levels (Watts, 1980; Watts and Stuiver, 1980; Watts and Hansen, 1988), which accelerated the development of karst. With the melting of the ice, sea levels and ground-water levels rose and many of the karst features were submerged. Examples of these flooded features include the "blue holes" found in the Bahamas, the cenotes of the Yucatan, the springs of Florida, and numerous water-filled cave passages throughout these terranes. Many of the numerous lakes and ponds of west-central Florida formed as overburden materials settled into cavities in the underlying limestone.

Mining exposed this typical karst limestone surface, which is riddled with dissolution cavities.

(William A. Wisner, 1972)

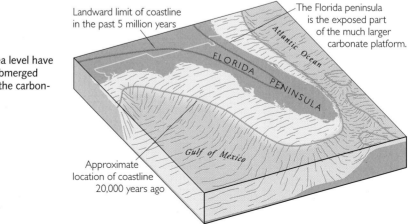

Changes in sea level have alternately submerged and exposed the carbonate platform.

Landward limit of coastline in the past 5 million years

The Florida peninsula is the exposed part of the much larger carbonate platform.

Atlantic Ocean

FLORIDA PENINSULA

Gulf of Mexico

Approximate location of coastline 20,000 years ago

Karst is an important part of the ground-water plumbing

At present, in west-central Florida, most of the soluble bedrock is below the water table. As ground water flows through the rock, geochemical processes continually modify both the rock and the chemical composition of the ground water. In many areas within the platform, the carbonates continue to dissolve, further enlarging cavities and conduits for ground-water flow. Fractures, faults, bedding planes and differences in the mineral composition of the carbonate rocks also play a role in the development, orientation, and extent of the internal plumbing system. Lineaments (linear features expressed in the regional surface terrain and often remotely sensed using aerial photography or satellite imagery) are often associated with locations of sinkholes and highly transmissive zones in the carbonate platform (Lattman and Parizek, 1964; Littlefield, and others, 1984).

THE MANTLED KARST OF WEST-CENTRAL FLORIDA

Where karst processes affect rocks that are covered by relatively insoluble deposits, the presence of buried karst features forms a distinctive type of terrain known as mantled karst. In mantled karst regions, the carbonate units are not exposed at land surface, but their presence may be indicated by sinkholes and the hummocky topography that results when the covering deposits take the shape of the underlying depressions. The mantled karst of west-central Florida has resulted in a number of distinct geomorphic regions (White, 1970; Brooks, 1981), including several lake districts with numerous lakes created by subsidence of overburden into the buried karst surface. In other areas, especially where the mantling deposits are thick, the buried karst surface is not reflected in the topography.

Sinkhole formation is related to the thickness and composition of the overlying materials

The mantled karst of west-central Florida has been classified into four distinct zones on the basis of the predominant type of sinkholes (Sinclair and Stewart, 1985). The type and frequency of sinkhole-subsidence activity have been correlated to the composition and thickness of overburden materials, the degree of dissolution within the underlying carbonate rocks, and local hydrologic conditions. Three general types of sinkholes occur: dissolution sinkholes—depressions in the limestone surface caused by chemical erosion of limestone; cover-subsidence sinkholes—formed as overburden materials gradually infill subsurface cavities; and cover-collapse sinkholes—also formed by movement of cover materials into subsurface voids, but characteristically formed more abruptly.

In the northern part of the region a thin (0 to 30 feet thick) mantle of highly permeable sediments overlies the carbonate rock. Rain water moves rapidly into the subsurface, dissolving the carbonate

In mantled karst terrane, the buried carbonate rock is furrowed and pitted. When the covering deposits subside into the underlying depressions, sinkholes and a hummocky topography result.

Overburden (mantle)

Carbonate bedrock (karst)

(Keith Bennett, Williams Earth Sciences Inc.)

The type, location, and frequency of sinkhole subsidence in the Southwest Florida Management District of west-central Florida have been related to the type and thickness of overburden materials.

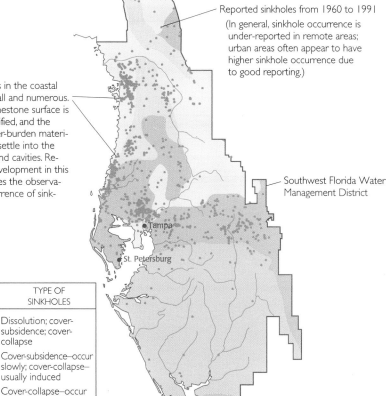

Reported sinkholes from 1960 to 1991
(In general, sinkhole occurrence is under-reported in remote areas; urban areas often appear to have higher sinkhole occurrence due to good reporting.)

New sinkholes in the coastal region are small and numerous. The buried limestone surface is intensely karstified, and the thin, sandy over-burden materials constantly settle into the buried voids and cavities. Recent urban development in this region increases the observation and occurrence of sinkhole activity.

Southwest Florida Water Management District

(Sinclair and Stewart, 1985; Wilson and Shock, 1996)

TYPE AND THICKNESS OF OVERBURDEN	FREQUENCY OF SINKHOLES	TYPE OF SINKHOLES
Thin; highly permeable	Generally few	Dissolution; cover-subsidence; cover-collapse
30 to 200 feet thick; permeable sands are dominant	Numerous	Cover-subsidence—occur slowly; cover-collapse—usually induced
30 to 200 feet thick; more clayey	Very numerous	Cover-collapse—occur abruptly
Greater than 200 feet	Few	Cover-collapse—large diameter and deep

A cover-collapse sinkhole formed in an orange grove east of Tampa.

rock, and dissolution-type sinkholes tend to develop. The slow dissolution of carbonates in these terranes has little direct impact on human activity (Culshaw and Waltham, 1987).

To the south, the overburden materials are generally thicker and less permeable. Where the overburden is 30 to 200 feet thick, sinkholes are numerous and two types are prevalent, cover-subsidence and cover-collapse. Where permeable sands are predominant in the overburden, cover-subsidence sinkholes may develop gradually as the sands move into underlying cavities. Where the overburden contains more clay, the greater cohesion of the clay postpones failure, and the ultimate collapse tends to occur more abruptly.

In the southernmost part of the region, overburden materials typically exceed 200 feet in thickness and consist of cohesive sediments interlayered with some carbonate rock units. Although sinkhole formation is uncommon under these geologic conditions, where sinkholes do occur they are usually large-diameter, deep, cover-collapse type.

Categorizing sinkholes
Two processes create three types of sinkholes

Three types of sinkholes are common in Florida: dissolution, cover-subsidence and cover-collapse sinkholes. They develop from dissolution and "suffosion." Dissolution is the ultimate cause of all sinkholes, but the type of sinkhole is also controlled by the thickness and type of overburden materials and the local hydrology.

Although it is convenient to divide sinkholes into three distinct types, sinkholes can be a combination of types or may form in several phases.

Cover-collapse sinkhole
near Ocala, Florida

(Tom Scott)

PROCESSES

Dissolution of soluble carbonate rocks by weakly acidic water is ultimately responsible for virtually all the sinkholes found in Florida.

ATMOSPHERE

Carbon dioxide (CO_2) Water (H_2O)

MANTLE or COVER SEDIMENT

Carbon dioxide (CO_2) Water (H_2O)

Carbonic acid (H_2CO_3)

CARBONATE BEDROCK (Limestone and dolomite)

Dolomite $[CaMg(CO_3)_3]$ Limestone $(CaCO_3)$

Magnesium, Calcium, Bicarbonate (Mg^{++}) (Ca^{++}) (HCO_3^-)

Water (H_2O) falling through the atmosphere and percolating the ground dissolves carbon dioxide (CO_2) gas from the air and soil, forming a weak acid—carbonic acid (H_2CO_3).

As the carbonic acid infiltrates the ground and contacts the bedrock surfaces, it reacts readily with limestone $(CaCO_3)$ and/or dolomite $[CaMg(CO_3)_3]$.

Cavities and voids develop as limestone or dolomite is dissolved into component ions of calcium (Ca^{++}), magnesium (Mg^{++}), and bicarbonate (HCO_3^-).

When the ground water becomes supersaturated with dissolved minerals, further dissolution is not possible, and carbonate salts of calcium and magnesium may precipitate from the water, often forming interesting shapes such as stalactites. The reactions are fully reversible, and when precipitates are exposed to undersaturated ground water they may redissolve. The geochemical interactions are controlled partly by the rate of circulation of water.

Suffosion occurs when unconsolidated overburden sediments infill preexisting cavities below them. This downward erosion of unconsolidated material into a preexisting cavity is also called raveling and describes both the catastrophic cover-collapse sinkhole and the more gradual cover-subsidence sinkhole.

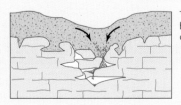

The erosion begins at the top of the carbonate bedrock and develops upward through the overlying sediments toward the land surface.

TYPES OF SINKHOLES

Dissolution of the limestone or dolomite is most intensive where the water first contacts the rock surface. Aggressive dissolution also occurs where flow is focussed in pre-existing openings in the rock, such as along joints, fractures, and bedding planes, and in the zone of water-table fluctuation where ground water is in contact with the atmosphere.

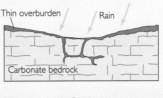

Rainfall and surface water percolate through joints in the limestone. Dissolved carbonate rock is carried away from the surface and a small depression gradually forms.

On exposed carbonate surfaces, a depression may focus surface drainage, accelerating the dissolution process. Debris carried into the developing sinkhole may plug the outflow, ponding water and creating wetlands.

Gently rolling hills and shallow depressions caused by solution sinkholes are common topographic features throughout much of Florida.

Cover-subsidence sinkholes tend to develop gradually where the covering sediments are permeable and contain sand.

Granular sediments spall into secondary openings in the underlying carbonate rocks.

A column of overlying sediments settles into the vacated spaces (a process termed "piping").

Dissolution and infilling continue, forming a noticable depression in the land surface.

The slow downward erosion eventually forms small surface depressions 1 inch to several feet in depth and diameter.

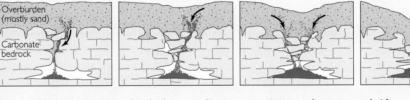

In areas where cover material is thicker or sediments contain more clay, cover-subsidence sinkholes are relatively uncommon, are smaller, and may go undetected for long periods.

Cover-collapse sinkholes may develop abruptly (over a period of hours) and cause catastrophic damages. They occur where the covering sediments contain a significant amount of clay.

Sediments spall into a cavity.

As spalling continues, the cohesive covering sediments form a structural arch.

The cavity migrates upward by progressive roof collapse.

The cavity eventually breaches the ground surface, creating sudden and dramatic sinkholes.

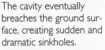

Over time, surface drainage, erosion, and deposition of sediment transform the steep-walled sinkhole into a shallower bowl-shaped depression.

Many of the numerous lakes and ponds that dot the Florida landscape, such as these in central Polk County, are actually subsidence depressions that are filled with water.

SINKHOLE DEVELOPMENT IS AFFECTED BY THE HYDROGEOLOGIC FRAMEWORK

The flow of subsurface water through sediments and eroded carbonate rocks affects how, where, and when sinkholes develop. Thus, formation of sinkholes is sensitive to changes in hydraulic and mechanical stresses that may occur naturally or as the result of human activity. Whether the stresses are imposed over geologic time scales by changes in sea level or over the time scale of human ground-water-resources development, they are expressed as changes in ground-water levels (hydraulic heads) and the gradients of hydraulic head. The hydraulic properties of the aquifers and the extent, composition, and thickness of overburden materials control how these stresses are transmitted. The chemistry of the ground water determines where dissolution and karst development occurs. Together, these hydrogeologic factors control the type and frequency of sinkholes that develop in west-central Florida.

Just as the hydrogeologic framework influences the development of sinkholes, the sinkholes influence the hydrogeologic framework. Understanding of the hydrogeologic framework can lead to land- and water-resources management strategies that minimize the impact of sinkholes.

Vast aquifer systems underlie west-central Florida

The hydrogeologic framework of west-central Florida consists of three layered aquifer systems that include both carbonate and siliciclastic rocks. The shallowest or "surficial" aquifer system generally occurs within unconsolidated sand, shell, and clay units. The surficial aquifer system ranges from less than 10 to more than 100 feet in thickness throughout west-central Florida. The water table is generally close to the land surface, intersecting lowlands, lakes, and streams. Recharge is primarily by rainfall. When sinkholes occur, it is the surficial aquifer deposits that commonly fail and move to infill any underlying cavities.

A sinkhole that breached a confining clay layer illustrates the interconnectivity of the aquifers. The water-level drop in the surficial aquifer system and the coincident rise in the Upper Floridan aquifer occurred as the sinkhole drained.

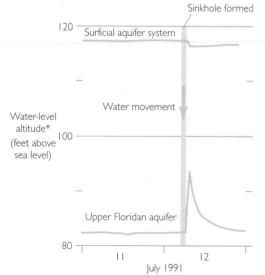

*Water levels were recorded at a SWFWMD Regional Observation Monitoring Program wellsite that is less than 1,000 feet from the sinkhole

(Southwest Florida Water Management District, written communication, 1998)

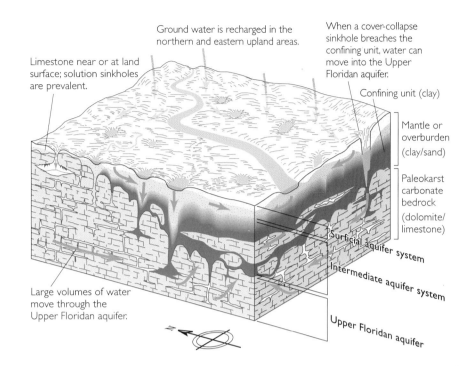

Ground water is recharged in the northern and eastern upland areas.

When a cover-collapse sinkhole breaches the confining unit, water can move into the Upper Floridan aquifer.

Limestone near or at land surface; solution sinkholes are prevalent.

Confining unit (clay)

Mantle or overburden (clay/sand)

Paleokarst carbonate bedrock (dolomite/limestone)

Surficial aquifer system

Intermediate aquifer system

Large volumes of water move through the Upper Floridan aquifer.

Upper Floridan aquifer

The type and frequency of sinkholes in west-central Florida are related to the presence or absence of the intermediate aquifer system.

In most of west-central Florida the surficial aquifer system is separated from the Upper Floridan aquifer by a hydrogeologic unit known as either the "intermediate aquifer system" or "intermediate confining unit," depending upon its local hydraulic properties (Southeastern Geological Society, 1986). The intermediate confining unit, delineated as such where fine-grained clastic deposits are incapable of yielding significant quantities of water, impedes the vertical flow of ground water between the overlying surficial aquifer system and the underlying Floridan aquifer system. In northern west-central Florida, where this unit is absent, the surficial aquifer system lies directly above the Floridan aquifer system. In general, the intermediate confining unit consists of heterogenous siliciclastic sediments that mantle the carbonate platform. These deposits thicken westward and southward, where they include more permeable clastic sediments and interbedded carbonate units. In these regions they are referred to as the intermediate aquifer system. The lateral extent of permeable units within the intermediate aquifer system is limited, and the transmissivities of these units are significantly smaller than those of underlying carbonate rocks of the Floridan aquifer system. The type and frequency of sinkholes in west-central Florida are correlated to the presence or absence of this intermediate layer and, where present, its composition and thickness.

The thick carbonate units of the Floridan aquifer constitute one of the most productive aquifer systems in the world. The Upper Floridan aquifer is between 500 and 1,800 feet thick and is the primary source of springflow and ground-water withdrawals in west-central Florida. Transmissivities commonly range from 50,000 to 500,000 square feet per day and may be as large as 13,000,000 square feet per day near large springs (Ryder, 1985). These transmis-

The presence of a confining unit affects the water level and the potential for sinkholes.

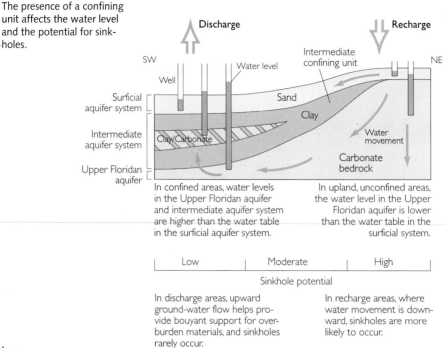

In confined areas, water levels in the Upper Floridan aquifer and intermediate aquifer system are higher than the water table in the surficial aquifer system.

In upland, unconfined areas, the water level in the Upper Floridan aquifer is lower than the water table in the surficial system.

In discharge areas, upward ground-water flow helps provide bouyant support for over-burden materials, and sinkholes rarely occur.

In recharge areas, where water movement is downward, sinkholes are more likely to occur.

Water in the Upper Floridan aquifer moves from recharge areas in the northern and eastern upland regions toward discharge areas near the coast.

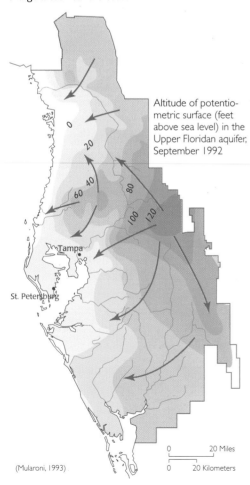

Altitude of potentiometric surface (feet above sea level) in the Upper Floridan aquifer, September 1992

(Mularoni, 1993)

sivity values far exceed those typical of diffuse ground-water flow in porous media such as sand and reflect the influence of karst-dissolution features.

In upland regions, hydraulic heads in the Upper Floridan aquifer are generally lower than heads in the surficial and intermediate aquifer systems. In these areas ground water moves downward from the surficial aquifer system, recharging the intermediate aquifer system and the Upper Floridan aquifer. This downward movement of ground water enhances the formation of sinkholes by facilitating raveling of unconsolidated sediments into the subterranean cavities. Where the intermediate confining unit is present, recharge to the Upper Floridan aquifer may be diminished. However, where the clay content of the confining unit is low, or the unit has been breached by sinkhole collapse or subsidence, downward movement of water and sediments from the surficial aquifer system can be greatly accelerated. Vertical shafts and sand-filled sinkholes can form high-permeability pathways through otherwise effective confining units (Brucker and others, 1972; Stewart and Parker, 1992).

Artesian conditions exist along much of the coast and, where confinement is poor, springs commonly occur. Parts of the northern coastal area are highly karstified, and the Upper Floridan aquifer is exposed at the land surface except where it is covered by unconsolidated sands. In the southern coastal regions, where the intermediate aquifer system and the Upper Floridan aquifer are well confined, water levels in those deeper units are higher than those in the surficial aquifer system, and ground water moves upward toward the surficial aquifer. Sinkholes rarely occur under these conditions.

Seasonal changes affect
ground-water levels
and sinkhole formation.

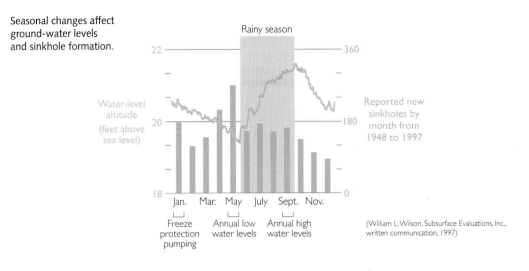

(William L. Wilson, Subsurface Evaluations, Inc.,
written communication, 1997)

Cyclical changes in water levels often occur in response to seasonal conditions in west-central Florida. At the end of the dry season, in May, ground-water levels are near their annual lows and, after the rainy season, in September, recover to their annual high levels. The range between the annual minimum and maximum levels can be significant. In some areas, especially during prolonged drought or large rainfall events, seasonal change in ground-water levels can lead to temporary reversals in the direction of vertical flow. More new sinkholes form during periods when ground-water levels are low.

Temporary reversals in head gradients may also be created by extreme, short-lived pumping. Longer-term ground-water pumping can lead to sustained ground-water level declines and gradient reversals, creating new recharge areas within the aquifer system and sometimes converting flowing springs to dry sinkholes. After the pumping stops, ambient conditions are usually restored, but the changes can become semipermanent or permanent if pumping persists over long periods of time, or confining units are compromised.

Long-term ground-water
pumping near Kissengen
Spring in central Polk County
led to a decline in water levels and ultimately caused the
spring to stop flowing.

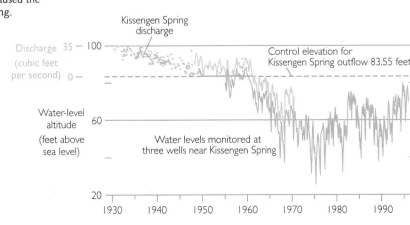

(Lewelling and others, 1998)

Changes in relative water levels caused by human activity can induce sinkholes

Normal conditions

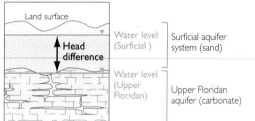

Land surface

Head difference

Water level (Surficial)   Surficial aquifer system (sand)

Water level (Upper Floridan)   Upper Floridan aquifer (carbonate)

Loading

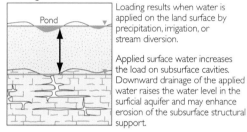

Pond

Loading results when water is applied on the land surface by precipitation, irrigation, or stream diversion.

Applied surface water increases the load on subsurface cavities. Downward drainage of the applied water raises the water level in the surficial aquifer and may enhance erosion of the subsurface structural support.

Pumping

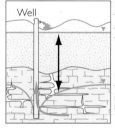

Well

Pumping commonly involves extraction of water from the lower aquifer and subsequent discharge onto the land surface.

Pumping may increase the gradient for downward drainage by increasing the head difference between the Upper Floridan and surficial aquifers.

Loading and pumping

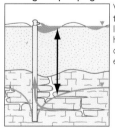

When loading and pumping occur together, the increased overburden load on subsurface cavities and enhanced downward drainage may combine to increase downward erosion or collapse cavities.

GROUND-WATER PUMPING, CONSTRUCTION, AND DEVELOPMENT PRACTICES INDUCE SINKHOLES

New sinkholes have been correlated to land-use practices (Newton, 1986). Induced sinkholes are conceptually divided into two types: those resulting from ground-water pumping (Sinclair, 1982) and those related to construction and development practices. Modified drainage and diverted surface water commonly accompany construction activities and can lead to focused infiltration of surface runoff, flooding, and erosion of sinkhole-prone earth materials. Manmade impoundments used to treat or store industrial- process water, sewage effluent, or runoff can also create a significant increase in the load bearing on the supporting geologic materials, causing sinkholes to form. Other construction activities that can induce sinkholes include the erection of structures, well drilling, dewatering foundations, and mining.

The overburden sediments that cover buried cavities in the aquifer systems are delicately balanced by ground-water fluid pressure. In sinkhole-prone areas, the lowering of ground-water levels, increasing the load at land surface, or some combination of the two may contribute to structural failure and cause sinkholes.

Aggressive pumping induces sinkholes

Aggressive pumping can induce sinkholes by abruptly changing ground-water levels and disturbing the equilibrium between a buried cavity and the overlying earth materials (Newton, 1986). Rapid declines in water levels can cause a loss of fluid-pressure support, bringing more weight to bear on the soils and rocks spanning buried voids. As the stresses on these supporting materials increase, the roof may fail and the cavity may collapse, partially filling with the overburden material.

Prior to water-level declines, incipient sinkholes are in a marginally stable stress equilibrium with the aquifer system. In addition to providing support, the presence of water increases the cohesion of sediments. When the water table is lowered, unconsolidated sediments may dry out and coarser-grained sediments, in particular, may move easily into openings.

Induced sinkholes are generally cover-collapse type sinkholes and tend to occur abruptly. They have been forming at increasing rates during the past several decades and pose potential hazards in developed and developing areas of west-central Florida. The increasing incidence of induced sinkholes is expected to continue as our demand for ground-water and land resources increases. Regional declines of ground-water levels increase sinkhole occurrence in sinkhole-prone regions. This becomes more apparent during the natural, recurring periods of low annual rainfall and drought.

Section 21 Well Field
Ground-water pumping for urban water supply induces new sinkholes

By the early 1930s, ground-water pumping along the west coast of Florida had lowered hydraulic heads in the fresh-water aquifers and caused upconing of saline water. Coastal municipalities began to abandon coastal ground-water sources and develop inland sources.

The city of St. Petersburg began pumping ground water from well fields in a rural area north of Tampa. By 1978, four well fields had been established in parts of Hillsborough, Pasco, and Pinellas Counties, and were pumping an average of 69,900 acre-feet per year. Sinkholes occurred in conjunction with the development of each of the well fields: Cosme (1930), Eldridge-Wilde (1954), Section 21 (1963), and South Pasco (1973).

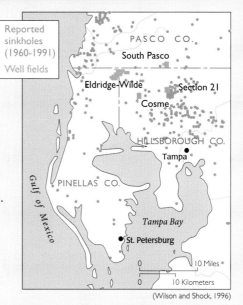

(Wilson and Shock, 1996)

SECTION 21 WELL FIELD

The effects of pumping on sinkhole development near the Section 21 well field illustrate the general relation between aggressive pumping, ground-water declines, and sinkhole development.

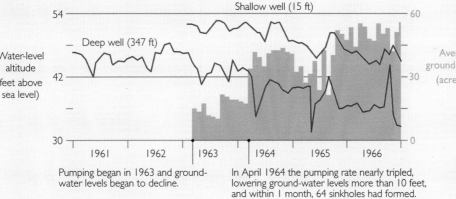

Pumping began in 1963 and ground-water levels began to decline.

In April 1964 the pumping rate nearly tripled, lowering ground-water levels more than 10 feet, and within 1 month, 64 sinkholes had formed.

(Sinclair, 1982)

Within 1 month of increasing the pumping rate, 64 new sinkholes formed within a 1-mile radius of the well field. Most of the sinkholes were formed in the vicinity of well 21-10, which was pumping at nearly twice the rate of the other wells. Neighboring areas also noticed dramatic declines in lake levels and dewatering of wetland areas.

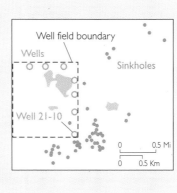

The sinkholes were apparently distributed randomly, except for those south and east of well 21-10, which were clustered along pre-existing joints.

The Section 21 well field is still in operation and researchers continue studying the effects of ground-water pumping on lake levels and wetlands.

(Sinclair, 1982)

Crop freeze protection
Heavy ground-water pumping during winter freezes produces new sinkholes

The mild winters are an important growing season for west-central Florida citrus, strawberry and nursery farmers. However, occasional freezing temperatures can result in substantial crop losses. To prevent freeze damage, growers pump warm (about 73° F) ground water from the Upper Floridan aquifer and spray it on plants to form an insulating coat of ice. Extended freezes have required intense and prolonged ground-water pumping, causing large drawdowns in the Upper Floridan aquifer and the abrupt appearance of sinkholes.

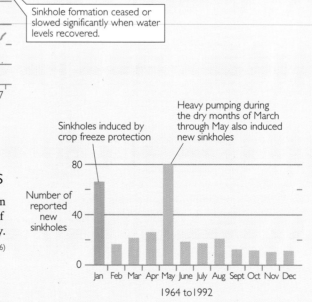

A thin layer of ice provides insulation from freezing temperatures.

(Tom Scott)

Pumping wells

Strawberry fields

Sinkholes

□ Dover

| Mi
| Km

In January 1977, extended freezes and associated ground-water withdrawals led to the sudden formation of 22 new sinkholes.

(Metcalf and Hall, 1984)

The relation between freezing weather, prolonged ground-water withdrawals, and sinkhole occurrence has been well documented in the Dover area about 10 miles east of Tampa (Bengtsson, 1987).

FREEZING AND PUMPING

During a 6-day period of record-breaking cold weather, ground water was pumped at night when temperatures fell below 39° F.

The new sinkholes were attributed to the movement of sandy overburden material through a breached clay confining unit into cavities in the limestone below.

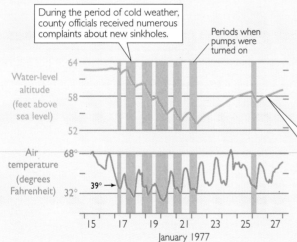

During the period of cold weather, county officials received numerous complaints about new sinkholes.

Periods when pumps were turned on

Water-level altitude (feet above sea level)

64

58

52

Air temperature (degrees Fahrenheit)

68°

39° →

32°

15   17   19   21   23   25   27
January 1977

Sinkhole formation ceased or slowed significantly when water levels recovered.

MANY NEW SINKHOLES

Ground-water pumping for crop freeze protection tends to induce sinkholes during the month of January in Hillsborough County.

(Wilson and Shock, 1996)

Sinkholes induced by crop freeze protection

Heavy pumping during the dry months of March through May also induced new sinkholes

Number of reported new sinkholes

80

40

0

Jan Feb Mar Apr May June July Aug Sept Oct Nov Dec

1964 to 1992

(Tom Scott)

"A giant sink hole opened up on Thursday, September 19 [1975] at a drilling site near Tampa, Florida and swallowed up a well-drilling rig, a water truck, and a trailer loaded with pipe all valued at $100,000. The well being drilled was down 200 ft when the ground began to give way to what turned out to be a limestone cavern. Within 10 minutes all the equipment was buried way out of sight in a crater measuring 300 ft deep, and 300 ft wide. Fortunately, the drilling crew had time to scramble to safety and no one was hurt."

—*from National Water Well Association newsletter*

One factor confounding the relation between pumping wells and the distribution of induced sinkholes is the nonuniform hydraulic connection between the well and various buried cavities. The development of secondary porosity is not uniform. Dissolution cavities often form along structural weaknesses in the limestone, such as bedding planes, joints, or fractures—places where water can more easily infiltrate the rock. The distribution of cavities can be controlled by the presence of these features and thus may be preferentially oriented. It is not uncommon for a pumping well to have more impact on cavities that are well-connected hydraulically—although farther away from the pumping well—than on nearby cavities that are less well-connected hydraulically. Proximity to pumping wells is not always a reliable indicator for predicting induced sinkholes.

When structures such as buildings and roadways are constructed, care is usually taken to divert surface-water drainage away from the foundations to avoid compromising their structural integrity. Associated activities may include grading slopes and removal or addition of vegetative cover, installing foundation piles and drainage systems, and ditching for storm drainages and conduits for service utilities. The altered landscapes typically result in local changes to established pathways of surface-water runoff, infiltration, and ground-water recharge. Pavements, roofs, and storm-drainage systems can dramatically increase the rate of ground-water recharge to a local area, thus increasing flow velocity in the bedrock and potentially inducing sinkholes. A common cause of induced sinkholes in urban areas is broken water or sewer pipes. Pipelines strung through karst terrane are subject to uneven settling as soils compact or are piped into dissolution cavities. The result can be cracked water pipes or the separation of sewer line sections, further aggravating erosion and perpetuating the process.

Loading by heavy equipment during construction or, later, by the weight of the structures themselves may induce sinkholes. A number of engineering methods are commonly used to prevent this type of sinkhole damage (Sowers, 1984), including drilling and driving pilings into competent limestone for support, injecting cement into subsurface cavities, and construction of reinforced and spread foundations that can span cavities and support the weight of the construction. Compaction by hammering, vibratory rollers, and heavy block drops may be used to induce collapse so that areas of weakness can be reinforced prior to construction.

"Construction practices often 'set the stage' for sinkhole occurrence."

—*J.G. Newton, 1986*

Excessive spray-effluent irrigation
Inducing sinkholes by surface loading

In April 1988 several cover-collapse sinkholes developed in an area where effluent from a wastewater treatment plant is sprayed for irrigation in northwestern Pinellas County. The likely cause was an increased load on the sediments at land surface due to waste-disposal activities, including periodic land spreading of dried sludge as well as spray irrigation. The 118-acre facility is located within a karst upland characterized by internal drainage and variable confinement between the surficial aquifer system and the Upper Floridan aquifer.

Spray-effluent volume applied for 1988 was equivalent to 290 inches per year (Trommer, 1992). Ponding of effluent occurred as the surficial sediments became saturated. The increased weight or load of the saturated sediments probably contributed to the ponding by causing some subsidence. At the beginning of the rainy season, several cover-collapse sinkholes developed suddenly, draining the effluent ponds into the aquifer system.

Sinkholes developed suddenly where water ponded due to excessive spray-effluent irrigation.   (John Trommer)

Heavy spraying of effluent raised water levels in the surficial aquifer system. As the ground became saturated, ponds formed.

The additional surface water, coupled with the onset of the rainy season, created strong potential for downward flow to Upper Floridan aquifer.

Several sinkholes developed, quickly draining the ponds.

Within days of sinkhole formation, discharge at Health Springs (at far left) increased dramatically.

Health Springs (north coast of Pinellas County)

Water table

Pond

Former pond level

West   Gulf of Mexico   East

Surficial aquifer system (sand)

Confining unit (clay)

Upper Floridan aquifer (limestone)

Not to scale

LINKING SURFACE AND GROUND WATER

Within several days of sinkhole formation, discharge at Health Springs, 2,500 feet downgradient in the ground-water flow path, increased from 2 cubic feet per second to 16 cubic feet per second (Trommer, 1992). Water-quality sampling of the spring during the higher flow detected constituents indicative of the spray effluent. Within 2 weeks, discharge at Health Springs had dropped to the normal rate of 2 cubic feet per second. The existence of a preferential ground-water flow path linking the upland spray field with the spring was confirmed by timing the movement of artificially dyed ground water between a well in the spray field and the spring (Tihansky and Trommer, 1994). The ground-water velocity

based on the arrival time of the dye was about 160 feet per day, or about 250 times greater than the estimates of the regional ground-water velocity (0.65 feet per day) in this area.

The dye-tracer test demonstrates how sinkholes and enhanced secondary porosity can provide a pathway directly linking surface-water runoff and the aquifer system. Sinkholes beneath holding ponds and rivers can convey surface waters directly to the Upper Floridan aquifer, and the introduction of contaminated surface waters through sinkholes can rapidly degrade ground-water resources.

Sinkhole collapse beneath a gypsum stack
Inducing sinkholes by surface loading and pumping

The sands and clays of the overburden sediments support a large phosphate mining and processing industry in west-central Florida. A gaping sinkhole formed abruptly on June 27, 1994, within a 400-acre, 220-foot high gypsum stack at a phosphate mine. The gypsum stack is a flat-topped pile of accumulated phosphogypsum—a byproduct of phosphate-ore chemical processing. The phosphogypsum precipitates when acidic mineralized water (about pH 1.5) used in processing the ore is circulated and evaporated from the top of the continually growing stack of waste gypsum. The waste slurry of slightly radioactive phosphogypsum results from the manufacture of phosphoric acid, a key ingredient in several forms of fertilizer.

The sinkhole likely formed from the collapse of a preexisting dissolution cavity that had developed in limestone deposits beneath the stack. Its development may have been accelerated by the aggressive chemical properties of the acidic waste slurry. Infiltration of the applied waste slurry into the underlying earth materials was unimpeded because there was no natural or engineered physical barrier immediately beneath the stack. Enlargement of cavities by dissolution and erosion combined with the increasing weight of the stack would have facilitated the sinkhole collapse. This effect may have been exacerbated by the reduction of fluid-pressure support for the overburden weight due to localized ground-water-level declines; the phosphate industry withdraws ground water from the Upper Floridan aquifer to supply water to the ore-refining plant.

The nearly vertical shaft tapered to a diameter of about 106 feet at a depth of 60 feet and extended more than 400 feet below the top of the stack.

An estimated 4 million cubic feet of phosphogypsum and an undetermined amount of contaminated water disappeared through the shaft.

(Hayward Baker, Inc., 1997)

Ground-water samples collected from the Upper Floridan aquifer confirmed that the aquifer had been locally contaminated with stack wastes. Officials began pumping nearby wells to capture the contaminated ground water and prevent its movement off-site.

PREVENTING SINKHOLE COLLAPSE

There are approximately 20 gypsum stacks located within the sinkhole-prone region of west-central Florida and, with the exception of new construction, all of these stacks are unlined. Because of potential environmental impacts from the phosphate industry, the State of Florida created the Phosphogypsum Management Rule to manage all aspects of phosphate chemical facilities. All new gypsum stacks are lined at their bases to impede the infiltration of process water and have specially designed water-circulation systems to prevent the escape of waste slurry. Ground-water-quality and water-level monitoring are also required. Efforts are being made to close all unlined stacks and reduce impacts on the underlying ground-water system. All new gypsum stacks must undergo an assessment of the susceptibility to subsidence activity and ground-water contamination. Geophysical surveys are used to locate potential zones of weakness so that any cavities or preexisting breaches can be plugged or avoided.

Before the collapse, acidic water was ponded on top of the stack to evaporate, leaving gypsum as a precipitate.

Acidic water percolated into the stack and ground-water system, thus accelerating development of the sinkhole.

Gypsum stack

Water level in stack

Sinkhole 160 ft

Rubble from the failed stack

220 ft

Land surface

Water movement

Mined surface

Sand (cast overburden)

Clay (confining unit, Hawthorn Formation)

Carbonate bedrock

Clay (confining unit)

Carbonate bedrock

Intermediate aquifer system

Upper Floridan aquifer system

Horizonal distance not to scale

A swarm of sinkholes suddenly appeared on a forest floor

Development of a new irrigation well triggered hundreds of sinkholes in a 6-hour period

Hundreds of sinkholes ranging in diameter from 1 foot to more than 150 feet formed within a 6-hour period on February 25, 1998, during the development of a newly drilled irrigation well (a procedure that involves flushing the well in order to obtain maximum production efficiency). Unconsolidated sand overburden collapsed into numerous cavities within an approximately 20-acre area as pumping and surging operations took place in the well.

Sinkholes induced during the development of an irrigation well affected a 20-acre area and ranged in size from less than 1 foot to more than 150 feet in diameter.

The affected land is located near the coast in an upland region that straddles parts of Pasco and Hernando counties. A 20-foot-thick sediment cover composed primarily of sand with little clay is underlain by cavernous limestone bedrock. The well was drilled through 140 feet of limestone, and a cavity was reported in the interval from 148 to 160 feet depth, where drilling was terminated. Very shortly after development began, two small sinkholes formed near the drill rig. As well development continued, additional new sinkholes of varying sizes began to appear throughout the area. Trees were uprooted and toppled as sediment collapse and slumping took place, and concentric extensional cracks and crevices formed throughout the landscape. The unconsolidated sandy material slumped and caved along the margins of the larger sinkholes as they continued to expand. The first two sinkholes to form eventually expanded to become the largest of the hundreds that formed during the 6-hour development period. They swallowed numerous 60-foot-tall pine trees and more than 20 acres of forest, and left the well standing on a small bridge of land.

TEST BORINGS AND HYDROGEOLOGIC DATA INDICATE SUSCEPTIBILITY TO SINKHOLES

The affected land contains several ponds formed by sinkholes long ago (paleosinkholes). Because west-central Florida is susceptible to sinkhole development, stability was tested along the margins of these ponds to determine if the site had higher-than-normal risks of sinkhole occurrence. Many test borings were made to measure the structural integrity of the bedrock, revealing a highly variable limestone surface. Two of the borings, approximately 100 feet apart, were made within a few hundred feet of the well site. One boring indicated that there was firm limestone at depth, whereas the other never encountered a firm foundation.

Irregularity in the limestone surface is typical of much of west-central Florida. Cavities, sudden bit drops, and lost circulation are frequently reported during drilling in this area. These drilling characteristics indicate the presence of significant cavernous porosity in the underlying limestone and, while commonly noted in drilling logs, only occasionally cause trouble during well construction.

Sinkhole susceptibility in this area is high

- The area is located within a mantled karst terrane where the limestone surface at depth is cavernous and highly irregular; the presence of nearby caves and springs suggests that major limestone dissolution has occurred.

- Water-level gradients are downward.

- Very little clay separates loose sand from limestone below.

- Previous sinkhole occurrence is well documented; the presence of paleosinkholes is evident on topographic maps of the region.

SINKHOLE IMPACTS CAN BE MINIMIZED

Sinkholes have very localized structural impacts, but they may have far-reaching effects on ground-water resources. Sinkholes can also impact surficial hydrologic systems—lakes, streams, and wetlands—by changing water chemistry and rates of recharge or run-off. Because the Earth's surface is constantly changing, sinkholes and other subsidence features will continue to occur in response to both natural and human-induced changes. We have seen how specific conditions can affect the type and frequency of sinkholes, including a general lowering of ground-water levels, reduced run-off, increased recharge, or significant surface loading. Recognition of these conditions is the first step in minimizing the impact of sinkholes.

In areas underlain by cavernous limestone with thin to moderate thickness of overburden, increased sinkhole development and property loss are strongly correlated to human activity and cultural development. There are several reasons for this correlation. First, rapid growth and development makes it more likely that new sinkholes will be reported, and the construction of roads and industrial or residential buildings increases exposure to the risk of property damage. Second, land-use changes in rapidly developing areas are often loosely controlled and include altered drainage, new impoundments for surface water, and new construction in sinkhole-prone areas. Finally, the changing land use is often associated with population increases and increasing demands for water supplies, which may lead to increases in ground-water pumpage and the lowering of local and regional ground-water levels.

Although we cannot adequately predict sinkhole development, we may be able to prevent or minimize the effects of sinkholes or reduce their rate of occurrence. Well-documented episodes of accelerated sinkhole activity are directly related to ground-water pumping events that lower ground-water levels. In many instances, the changes in ground-water levels are only a few tens of feet. It is

Cover-collapse sinkhole
Winter Park, 1981

(Tom Scott)

A newly formed sinkhole 20 miles north of Tampa is being examined by a team of scientists.

likely that many induced sinkholes can be prevented by controlling fluctuations in ground-water levels.

The overall regional decline in water levels in the Upper Floridan aquifer has been a long-standing concern of water-resource managers. Local declines around municipal well fields, often much greater than the regional declines, have led to dewatering of lakes and wetlands, upconing of poorer-quality water, saltwater intrusion, and accelerated sinkhole development. The Southwest Florida Water Management District has been working with other water-resources agencies to establish critical levels for ground water within the west-central Florida area. The establishment of minimum ground-water levels will help minimize sinkhole impacts by ameliorating some of the conditions that cause them.

Land-use planners, resource managers, and actuaries have been able to estimate the probability of sinkhole occurrence and associated risks. The Florida Department of Insurance designed insurance premiums for four sinkhole probability zones (Wilson and Shock, 1996) on the basis of insurance claims for sinkhole damages and hydrogeologic conditions. West-central Florida was delineated as an area having the highest frequency of sinkhole activity. The use of scientific information to assess risks and establish insurance rates demonstrates the benefits of understanding the hydrogeologic framework and potential effects of water-resource development. This scientific understanding is key to assigning meaningful risks to both property and the environment, and essential for formulating effective land- and water-resources management strategies.

THE ROLE OF SCIENCE

Land Subsidence in the United States

Scientists use equations that represent physical and chemical processes to analyze subsidence.

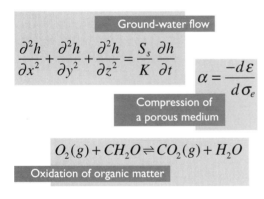

Ground-water flow

$$\frac{\partial^2 h}{\partial x^2} + \frac{\partial^2 h}{\partial y^2} + \frac{\partial^2 h}{\partial z^2} = \frac{S_s}{K} \frac{\partial h}{\partial t}$$

$$\alpha = \frac{-d\varepsilon}{d\sigma_e}$$

Compression of a porous medium

$$O_2(g) + CH_2O \rightleftharpoons CO_2(g) + H_2O$$

Oxidation of organic matter

Devin Galloway, S.E. Ingebritsen, and
Francis S. Riley
U.S. Geological Survey,
Menlo Park, California

Marti E. Ikehara
National Geodetic Survey,
Sacramento, California

Michael C. Carpenter
U.S. Geological Survey,
Tucson, Arizona

The Panel on Land Subsidence of the U.S. National Research Council (NRC) (1991) recognized three information needs: *"First, basic earth-science data and information on the magnitude and distribution of subsidence [...] to* **recognize** *and to* **assess** *future problems. These data […] help not only to address local subsidence problems but to identify national problems. [...] Second,* **research on subsidence processes** *and engineering methods for dealing with subsidence […] for cost-effective damage prevention or control. […] And third, although many types of* **mitigation** *methods are in use in the United States, studies of their cost-effectiveness would facilitate choices by decision makers."* (emphases added)

The third need can only be met after we learn how to better measure the total impact of subsidence problems and the effectiveness of our attempted solutions. It is clear that in order to assess the total impact we would need to inventory the total costs to society of overdrafting susceptible aquifer systems. Presently this is impractical because there are only sparse estimates of subsidence costs, and most of these are directly related to damages to tangible property. Additional consideration could be given to many of the indirect costs of excessive ground-water withdrawal and subsidence. In particular, it is our impression that the impact of subsidence on our surface-water resources and drainage—riparian and wetland habitat, drainage infrastructure, and flood risk—is large. Though much knowledge could be gained from risk-benefit analyses that include the indirect costs of subsidence, in this concluding chapter we focus on the role of science as identified by the NRC panel—recognition and assessment of subsidence, research on subsidence processes, and mitigation methods.

RECOGNITION

The occurrence of land subsidence is seldom as obvious as it is in the case of catastrophic sinkholes such as those in Winter Park, Florida, or at the Retsof Salt Mine in Genesee Valley, New York. Discovery of such catastrophic subsidence is difficult only when the localized collapse occurs in a remote area. Where ground-water mining or drainage of organic soils are involved, the subsidence is typically gradual and widespread, and its discovery becomes an exercise in detection. Gazing out over the San Joaquin Valley, California, one would be hard-pressed to recognize that more than 30 feet of subsidence has occurred in some locations. In the absence of obvious clues such as protruding wells, failed well casings, broken pipelines, and drainage reversals, repeat measurements of land-surface elevation are needed to reveal the subsidence.

The problem of detection in regional land subsidence is compounded by the large areal scale of the elevation changes and the requirement for vertically stable reference marks—bench marks—located outside the area affected by subsidence. Where such stable bench marks exist and repeat surveys are made, subsidence is fairly easily measured using professional surveying instruments and methods. In fact, this is one of the common ways in which subsidence is first detected. Often, public agencies or private contractors discover that key local bench marks have moved only after repeat surveys that span several years or longer. Prior to the discovery, when the cumulative subsidence magnitude is small, the apparent errors in the surveys may be adjusted throughout the network under the assumption that the discrepancies reflect random errors of the particular survey. The subsidence may then go undetected until later routine surveys, or until suspicions arise and steps are taken to confirm the current elevations of the affected bench marks.

Subsidence is sometimes obvious

Protruding well casings are common in agricultural areas and some urban areas where ground water has been extracted from alluvial aquifer systems. The land surface and aquifer system are displaced downward relative to the well casing, which is generally anchored at a depth where there is less compaction. The stressed well casings are subject to failure through collapse and dislocation. Submersible pumps, pump columns, and the well itself may be damaged or require rehabilitation. Deep wells are most vulnerable and are also the most expensive to repair and replace. Typical repair costs amount to $5,000–$25,000 or more, and replacement costs are in the range of $40,000–$250,000! Where the frequency of well-casing failures is high, land subsidence is often suspected and is often the cause.

The formation of earth fissures in alluvial aquifer systems is another indication that compaction and land subsidence may be occurring. Other possible indicators of land subsidence include changes in flood-inundation frequency and distribution; stagnation or reversals of streams, aqueducts, storm drainages, or sewer lines; failure, overtopping or reduction in freeboard along reaches of levees, canals, and flood-conveyance structures; and, more generally, cracks and/or changes in the gradient of linear engineered structures such as pipelines and roadways.

Well head

Subsidence

(Pictured is Terry Katzer)

An abandoned water-supply well protrudes above ground in Las Vegas, 1997.

Drought conditions in the Sacramento Valley during 1976-77 reduced the amount of surface water available for irrigation and, for the first time, more ground water than surface water was used to irrigate crops.

During the summer of 1977 many irrigation wells that penetrated the valley-fill deposits were damaged. Most of the damaged wells occurred in the southwestern part of the valley. The damage seems to have been caused by compaction of the aquifer system which resulted in the vertical compression and rupture of well casings.

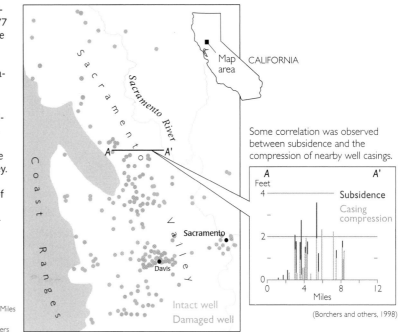

Some correlation was observed between subsidence and the compression of nearby well casings.

(Borchers and others, 1998)

This bench mark was established in 1995 to monitor potential land subsidence in the Antelope Valley, Mojave Desert, California.

ASSESSMENT

Differential surveys measure relative changes in the position of the land surface. The observable position is typically a geodetic mark that has been established to some depth (usually greater than 10 feet when in soil), so that any movement can be attributed to deep-seated ground movement and not to surficial effects such as frost heave. Sometimes geodetic marks, especially those used to measure vertical movement (bench marks), are established in massive artificial foundations, such as bridge abutments, that are well-coupled to the earth. Any vertical or horizontal movement of a geodetic mark is measured in relation to other observation points. When the bench mark can be assumed to be stable or its movement is otherwise known and measurable, it can be used as a control point, and the absolute position of the observation point can be determined. By this method, land subsidence has been measured using repeat surveys of bench marks referenced to some known, or presumed stable, reference frame. Access to a stable reference frame is essential for the measurements needed to map land subsidence. In many areas where subsidence has been recognized, and other areas where subsidence has not yet been well documented, accurate assessment has been hindered or delayed by the lack of a sufficiently stable vertical reference frame (control).

Known positions are linked into a network

"Sufficiently stable" is a somewhat relative term that has meaning in the context of a particular time-frame of interest and magnitude of differential movement. Because of continuous and episodic crustal motions caused mostly by postglacial rebound, tectonism, volcanism, and anthropogenic alteration of the Earth's surface, it is occasionally necessary to remeasure geodetic control on a national scale. Networks of geodetic control consist of known positions that are determined relative to a horizontal or vertical datum or both.

Two reference networks are used for horizontal and vertical geodetic control for the United States, the North American Datum of 1983 (NAD83) and the North American Vertical Datum of 1988 (NAVD88). NAD83 replaces the older North American Datum of 1927 (NAD27) and is the current geodetic reference system for horizontal control in the United States, Canada, Mexico, and Central America. It is the legally recognized horizontal control datum for the Federal government of the United States and for 44 of the 50 individual States. NAVD88 replaces the National Geodetic Vertical Datum of 1929 (NGVD 1929), which was based on local mean sea levels determined at 26 tidal gauges. The principal sea-level reference for NAVD88 is the primary tidal gauge at Father Point/Rimouski, Quebec, Canada. The vertical datum is based on the Earth's geoid—a measurable and calculable surface that is equivalent to mean sea level.

For more information on geodetic control, visit the National Geodetic Survey web site at **http://www.ngs.noaa.gov/faq.shtml**

In partnership with other public and private parties, the National Geodetic Survey (NGS) has implemented High Accuracy Reference Networks (HARNs) in every State. HARN observation campaigns (originating in Tennessee in 1986 and ending in Indiana in 1997) resulted in the establishment of some 16,000 survey stations. The updated networks were needed not only to replace thousands of historic bench marks and horizontal-control marks lost to development, vandalism, and natural causes, but also to provide geodetic monuments easily accessible by roadways. These updated reference networks will facilitate the early and accurate detection and measurement of land subsidence.

Spirit leveling was once a common method of determining elevation

Before the advent of the satellite-based Global Positioning System (GPS) in the 1980s, the most common means of conducting land surveys involved either the theodolite or, since the 1950s, the geodimeter (an electronic distance measuring device, or EDM). If only vertical position were sought, the spirit level has been the instrument of choice. The technique of differential leveling allows the surveyor to carry an elevation from a known reference point to other points by use of a precisely leveled telescope and graduated vertical rods. Despite its simplicity, this method can be very accurate. When surveying to meet the standards set for even the lower orders of accuracy in geodetic leveling, 0.05-foot changes in elevation can be routinely measured over distances of miles. At large scales, leveling and EDM measurement errors increase. When

USGS survey party spirit leveling near Colusa, Sacramento Valley, in 1904 and their field notes.

(John Ryan, donated courtesy of Thomas E. Ryan)

the scale of the survey is small (on the order of 5 miles or less) and the desired spatial density is high, spirit leveling is still commonly used because it is accurate and relatively inexpensive. Large regional networks warrant use of the more efficient Global Positioning System (GPS) surveying for differential surveys.

GPS—Global Positioning System uses Earth-orbiting satellites

A revolution in surveying and measurement of crustal motion occurred in the early 1980s when tests of the satellite-based NAVSTAR GPS showed that it was possible to obtain 1 part in 1 million precision between points spaced from 5 to more than 25 miles apart. GPS uses Earth-orbiting satellites to trilaterate positions based on the time required for radio signals transmitted from satellites to reach a receiving antenna. An accurate three-dimensional position can be determined from trilateration of the range distances between the receiver and at least four satellites. Since July 17, 1995, NAVSTAR has been operational with a full constellation of 24 satellites, and in North America provides essentially continuous coverage with at least 6 satellites in view. Guidelines have been formulated for establishing GPS-derived ellipsoid heights with accuracy standards at either the 2-cm (.0656 ft) or the 5-cm (.164 ft) level (Zilkoski and others, 1997).

In land-subsidence and other crustal-motion surveys, the relative positions of two points can be determined when two GPS receivers, one at each observation point, receive signals simultaneously from the same set of 4 or more satellites. When the same points are reoccupied following some time interval, any relative motion between the points that occurred during the time interval can be measured. Geodetic networks of points can be surveyed in this fashion. Such a network, one of the first of its kind designed specifically to monitor land subsidence, was established in the Antelope Valley, Mojave

A full constellation of the Global Positioning System (GPS) includes 24 satellites in orbit 12,500 miles above the Earth. The satellites are positioned so that we can receive signals from six of them at any one time from any point on the Earth.

(Jay Prendergast, 1992)

A GPS antenna mounted on a tripod at a known distance above a geodetic mark near Monterey, California, receives signals from GPS satellites. The operator is entering station information into a receiver that stores the signals for later processing.

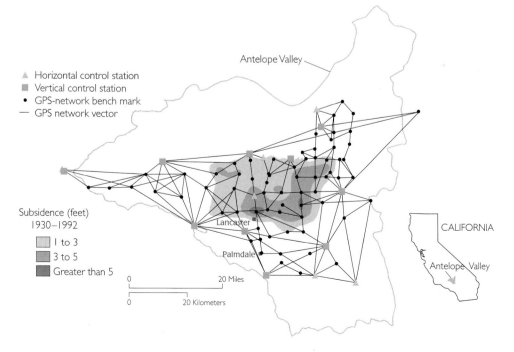

Horizontal control station
Vertical control station
GPS-network bench mark
GPS network vector

Antelope Valley

Lancaster

Palmdale

Subsidence (feet)
1930–1992

1 to 3
3 to 5
Greater than 5

0   20 Miles

0   20 Kilometers

CALIFORNIA

Antelope Valley

This geodetic network was used to measure historical subsidence in Antelope Valley, Mojave Desert, California.

Geodetic surveying of 85 stations in Antelope Valley using GPS required about 150 person-days during 35 days of observation in 1992. Results from the GPS surveys and conventional leveling surveys spanning more than 60 years showed a maximum subsidence of about 6.6 feet; more than 200 square miles had subsided more than 2 feet since about 1930 (Ikehara and Phillips, 1994).

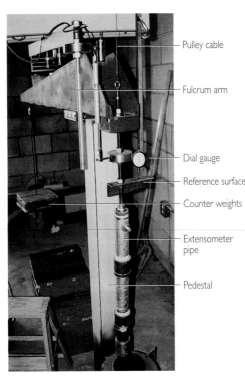

Part of a two-stage counter-weighted pipe extensometer that measures compaction in a shallow aquifer near Lancaster,

Desert, California in 1992 (Ikehara and Phillips, 1994). It was designed to determine the subsidence of previously leveled bench marks and enable precise measurement of points separated by tens of miles for future subsidence monitoring. Other large GPS-based geodetic networks for subsidence monitoring have been established in Albuquerque, New Mexico; the Avra Valley, Arizona; Las Vegas, Nevada; the Lower Coachella Valley, California; the Sacramento-San Joaquin Delta, California; and the Tucson basin, Arizona. GPS surveying is also a versatile exploratory tool that can be used in a rapid mode to quickly but coarsely define subsidence regions, in order to site more precise, site-specific and time-continuous measurement devices such as extensometers.

Extensometers measure subsidence and horizontal displacement

Borehole extensometers generate a continuous record of change in vertical distance between the land surface and a reference point or "subsurface bench mark" at the bottom of a deep borehole (Riley, 1986). In areas undergoing aquifer-system compaction, the extensometer is the most effective means of determining precise, continuous deformation at a point. If the subsurface bench mark is established below the base of the compacting aquifer system, the extensometer can be used as the stable reference or starting point for local geodetic surveys. Designs that incorporate multiple-stage extensometers in a single instrument are being used to measure aquifer-system compaction simultaneously in different depth intervals.

This two-stage, counter-weighted pipe extensometer measures compaction simultaneously in shallow and deep aquifers in Antelope Valley, Mojave Desert, California.

As the aquifer system compresses, the land surface subsides along with the extensometer table. The extensometer pipe anchored deeper in the aquifer system appears to rise relative to the table. This relative movement represents the amount of vertical displacement occurring in the aquifer system between the shallow-seated piers supporting the table and the bottom of the extensometer pipe.

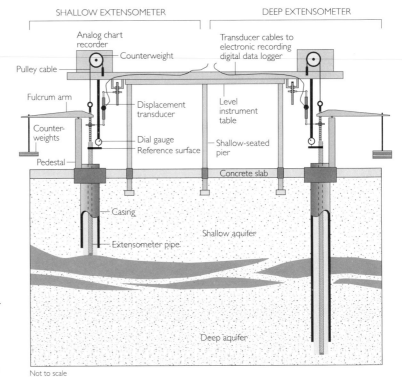

Measurements of water level and compaction from borehole extensometers form the basis of stress-strain analysis. These data are from the Hueco Basin, El Paso, Texas.

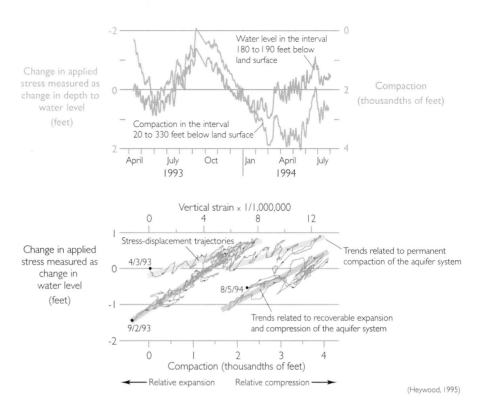

This quartz-tube horizontal extensometer is being installed near Apache Junction, Arizona. The quartz tube is placed inside the pipe housing and attached to a post on one side of a fissure. A displacement sensor, such as a dial gage, is attached to the post on the opposite side of the fissure and pushes against the quartz tube. Fissure opening and closing is observed in the dial-gage reading. An electronic sensor can be substituted for the dial gage for continuous measurement of fissure movement.

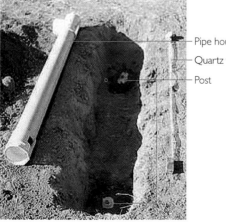

As a stand-alone instrument, the borehole extensometer may be regarded simply as a sentinel against the undetected onset of unacceptable rates of aquifer-system compaction. However, when used in conjunction with good well logs and water-level data from an adjacent observation well, the deformation history generated by an extensometer can provide the basis for stress-strain analysis (Riley, 1969) and inverse modeling that defines the average compressibility and vertical hydraulic conductivity of the aquitards (Helm, 1975). This capability derives from the fact that the compaction measured by the extensometer is directly related to the volume of water produced by the aquitards. Major improvements in stability and sensitivity allow recently constructed extensometers to record the minute elastic compression and expansion that inevitably accompany even very small fluctuations in ground-water levels in unconsolidated alluvial aquifer systems, as well as the relatively large deformations typical of the irreversible compaction of aquitards. Reliable estimates of aquitard properties are necessary for constraining predictive modeling, whether the objective is the prevention or mitigation of land subsidence or simply the optimal use of the storage capacity of the aquifer system.

Several kinds of horizontal extensometers measure differential horizontal ground motion at earth fissures caused by changes in ground-water levels (Carpenter, 1993). Buried horizontal extensometers constructed of quartz tubes or invar wires are useful when precise, continuous measurements are required on a scale of 10 to 100 feet. Tape extensometers measure changes across intermonument distances up to 100 feet with a repeatability of 0.01 inches. The tape

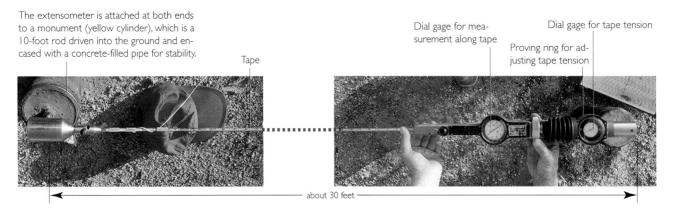

The extensometer is attached at both ends to a monument (yellow cylinder), which is a 10-foot rod driven into the ground and encased with a concrete-filled pipe for stability.

Tape

Dial gage for measurement along tape

Dial gage for tape tension

Proving ring for adjusting tape tension

about 30 feet

Tape extensometers measure horizontal ground motion over distances of up to 100 feet.

extensometer is used in conjunction with geodetic monuments specially equipped with ball-bearing instrument mounts, which can serve as both horizontal and vertical control points. Arrays or lines of monuments can be extended for arbitrary distances, usually in the range of 200 to 600 feet.

Radar interferometry is a new tool for measuring subsidence

Interferometric Synthetic Aperture Radar (InSAR) is a powerful new tool that uses radar signals to measure deformation of the Earth's crust at an unprecedented level of spatial detail and high degree of measurement resolution. Geophysical applications of radar interferometry take advantage of the phase component of reflected radar signals to measure apparent changes in the range distance of the land surface (Gabriel and others, 1989; Massonnet

Radar* is an active sensor, transmitting a signal of electromagnetic energy. Satellite-borne radar using one antenna transmits a pulsed train of microwaves.

The waves reflect off the ground surface, and echoes are received by the moving antenna, producing a recorded image of the scanned ground that is continuous along the track of the satellite and about 60 miles wide.

The restricted size of the satellite antenna limits the spatial resolution to 3 to 6 miles on the ground.

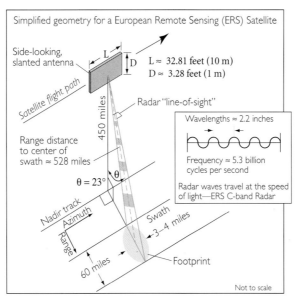

Simplified geometry for a European Remote Sensing (ERS) Satellite

Side-looking, slanted antenna

Satellite flight path

L ≈ 32.81 feet (10 m)
D ≈ 3.28 feet (1 m)

Radar "line-of-sight"

450 miles

Range distance to center of swath ≈ 528 miles

Wavelengths ≈ 2.2 inches

Frequency ≈ 5.3 billion cycles per second

Radar waves travel at the speed of light—ERS C-band Radar

θ = 23°

Nadir track

Azimuth

Range

Swath 3–4 miles

60 miles

Footprint

Not to scale

*RADAR: RAdio Detection And Ranging

Synthetic Aperture Radar (SAR) imaging can "synthesize" an effectively larger antenna (about 3 miles long) with a spatial resolution on the order of 16 feet by pulsing the microwaves every 16 feet of satellite travel.

The 3- to 4-mile-wide footprints overlapped at 16-foot intervals along the ground track are processed through a technique similar to medical x-ray imaging. Numerous 16-foot echoes are averaged to improve signal coherence, and the actual spatial resolution is on the order of 260 feet or better.

Interferograms are made by differencing successive SAR images taken from the same orbital position but at different times. Under favorable radiometric conditions 1/2-inch to 1/10-inch resolution is possible in the line-of-sight (range) of the radar.

and Feigl, 1998). Ordinary radar on a typical Earth-orbiting satellite has a very poor ground resolution of about 3–4 miles because of the restricted size of the antenna on the satellite. Synthetic Aperture Radar (SAR) takes advantage of the motion of the spacecraft along its orbital track to mathematically reconstruct (synthesize) an operationally larger antenna and yield high-spatial-resolution imaging capability on the order of tens of feet. The size of a picture element (pixel) on a typical SAR image made from satellite-borne radar may be as small as 1,000 square feet or as large as 100,000 square feet, depending how the image is processed.

For landscapes with more or less stable radar reflectors (such as buildings or other engineered structures, or undisturbed rocks and ground surfaces) over a period of time, it is possible to make high-precision measurements of the change in the position of the reflectors by subtracting or "interfering" two radar scans made of the same area at different times. This is the principle behind InSAR.

Under ideal conditions, it is possible to resolve changes in elevation on the order of 0.4 inches (10 mm) or less at the scale of 1 pixel. Interferograms, formed from patterns of interference between the phase components of two radar scans made from nearly the same antenna position (viewing angle) but at different times, have demonstrated dramatic potential for high-density spatial mapping of ground-surface deformations associated with tectonic (Massonnet and others, 1993; Zebker and others, 1994) and volcanic strains (Massonnet and others, 1995; Rosen and others, 1996; Wicks and

Differential interferogram made from InSAR images acquired June 1993 and March 1997 over the Tucson and Green Valley areas of Arizona. The interferogram (center) is shown overlain on the radar amplitude image.

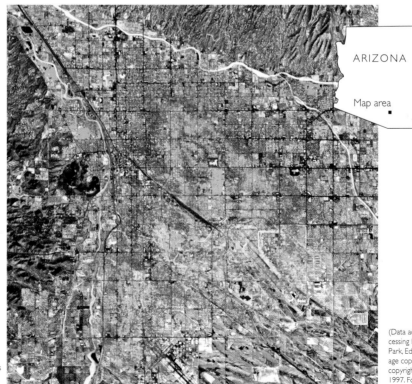

One color cycle represents about 1.1 inch of subsidence. More than two cycles can be seen on this Tucson image.

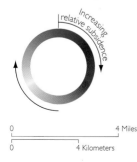

(Data acquisition and interferometric processing by the NPA Group, Crockham Park, Edenbridge, Kent TN8 6SR, UK. Image copyright: NPA 1998. Image data copyright: European Space Agency 1993, 1997. For information contact ren@npagroup.co.uk)

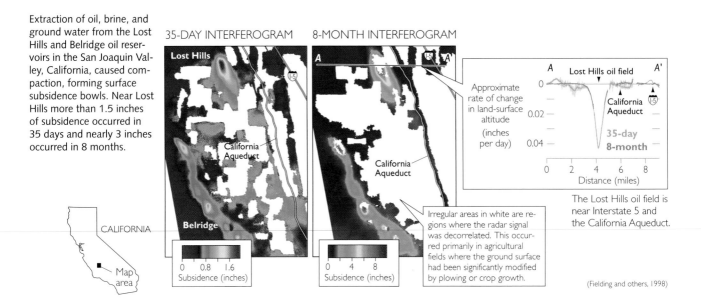

Extraction of oil, brine, and ground water from the Lost Hills and Belridge oil reservoirs in the San Joaquin Valley, California, caused compaction, forming surface subsidence bowls. Near Lost Hills more than 1.5 inches of subsidence occurred in 35 days and nearly 3 inches occurred in 8 months.

35-DAY INTERFEROGRAM   8-MONTH INTERFEROGRAM

Approximate rate of change in land-surface altitude (inches per day)

The Lost Hills oil field is near Interstate 5 and the California Aqueduct.

Irregular areas in white are regions where the radar signal was decorrelated. This occurred primarily in agricultural fields where the ground surface had been significantly modified by plowing or crop growth.

(Fielding and others, 1998)

others, 1998). InSAR has also recently been used to map localized crustal deformation and land subsidence associated with geothermal fields in Imperial Valley, California (Massonnet and others, 1997), Long Valley, California (W. Thatcher, USGS, written communication, 1997), and Iceland (Vadon and Sigmundsson, 1997), and with oil and gas fields in the Central Valley, California (Fielding and others, 1998). InSAR has also been used to map regional-scale land subsidence caused by aquifer-system compaction in the Antelope Valley, California (Galloway and others, 1998), Las Vegas Valley, Ne-

Different methods of measuring land subsidence

METHOD	Component displacement	Resolution[1] (milimeters)	Spatial density[2] (samples/survey)	Spatial scale (elements)
Spirit level	vertical	0.1–1	10–100	line-network
Geodimeter	horizontal	1	10–100	line-network
Borehole extensometer	vertical	0.01–0.1	1–3	point
Horizontal extensometer:				
Tape	horizontal	0.3	1–10	line-array
Invar wire	horizontal	0.0001	1	line
Quartz tube	horizontal	0.00001	1	line
GPS	vertical horizontal	20 5	10–100	network
InSAR	range	10	100,000– 10,000,000	map pixel[3]

[1]Measurement resolution attainable under optimum conditions. Values are given in metric units to conform with standard geodetic guidelines. (One inch is equal to 25.4 millimeters and 1 foot is equal to 304.8 millimeters.)

[2]Number of measurements generally necessary to define the distribution and magnitude of land subsidence at the scale of the survey.

[3]A pixel on an InSAR displacement map is typically 40 to 80 meters square on the ground.

vada (Amelung and others, 1999), and Santa Clara Valley, California (Ikehara and others, 1998).

RESEARCH ON SUBSIDENCE PROCESSES

The areal and vertical distribution of subsidence-prone materials, their current state of stress, and their stress history govern the potential for subsidence. These factors vary in importance and can be determined with varying degrees of difficulty for the three major types of subsidence considered in this Circular.

In the case of organic-soil subsidence (oxidation), the subsidence-prone material is generally surficial, and both thickness and areal extent are often readily mapped. The state of stress and the stress history are largely irrelevant, as the subsidence rate is mainly determined by the degree of drainage. In aquifer-system compaction, the subsidence-prone (fine-grained) material is buried and must be mapped indirectly by drilling, sampling, assembling drilling logs of the subsurface lithology, and by various borehole and surface geophysical techniques. These methods produce spatially discrete information—often one-dimensional or, in the case of surface geophysics, quasi-two dimensional with integrated depth information. The interpretation is often ambiguous and extrapolation of the spatially limited data to other areas of interest is laden with uncertainty, making the mapping imperfect. Mapping of subsidence-prone materials is perhaps most difficult for those materials subject to catastrophic collapse, because the failures are so localized and frequently evolve over short time scales. Acoustic profiling has been used successfully to map possible locations of buried cavities in west-central Florida. For both aquifer-system compaction and cavity collapse, the current stress and stress history are critically important.

An acoustic profile taken across a sinkhole lake in west-central Florida shows how high-resolution seismic-reflection techniques can image geologic characteristics associated with subsidence.

(Tihansky and others, 1996)

Water

Lake bottom

Recent lake sediments

Intact sedimentary structure with little disruption

50

Feet

0   400

Units have failed or have been disrupted, leaving little identifiable geologic structure.

Dipping reflectors along margins of buried cavities where geologic units are deformed and are sagging into underlying void space

In typical alluvial ground-water basins, an accurate initial estimate of preconsolidation stress (critical head) is particularly important for successful evaluation of the historical compaction of aquifer systems (Hanson, 1989; Hanson and Benedict, 1994). Prior to the development of ground-water resources in a basin, the alluvial sediments are typically overconsolidated—the initial preconsolidation stress of the aquifer system is larger than the intergranular or effective stresses. Land subsidence becomes obvious only after substantial water-level drawdowns have caused increased intergranular stresses and initiated inelastic compaction. Holzer (1981) identified a variety of natural mechanisms that can cause such an overconsolidated condition in alluvial basins, including removal of overburden by erosion, prehistoric ground-water-level declines, desiccation, and diagenesis. Few investigations have examined the elastic responses of the aquifer system to changes in effective stress under natural conditions, before large-scale ground-water withdrawal has begun to cause irreversible subsidence. As a result, information on critical aquifer hydraulic head, representing the native preconsolidation stress of the system, is usually deduced from paired time-series of ground-water levels and land subsidence (Holzer, 1981; Anderson, 1988, 1989) measured at wells and nearby bench marks, or inferred from ground-water-flow models (Hanson and others, 1990; Hanson and Benedict, 1994).

Similar uncertainties exist for systems that have undergone some period of lowered ground-water levels and land subsidence followed by ground-water-level recovery and slowing or cessation of subsidence. The problem of determining the new preconsolidation stress thresholds in these aquifer systems is equally as difficult as determining the native preconsolidation stresses in undeveloped aquifer systems. The difficulty is compounded when the developed aquifer systems contain thick aquitards affected by hydrodynamic lag.

Preconsolidation stress is usually deduced from paired time-series of ground-water levels and subsidence. Estimates of preconsolidation stress such as these from Antelope Valley, Mojave Desert, California, are used to evaluate the historical compaction of aquifer systems.

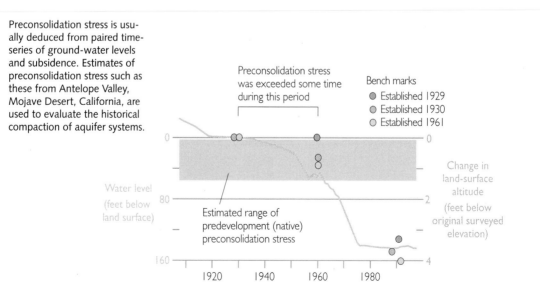

This three-layer digital model of Antelope Valley, Mojave Desert, California is a mathematical representation of the physical processes of ground-water flow and aquifer-system compaction. Separate model layers represent different depth horizons within the aquifer system. Flow and compaction properties are specified for each of the 2,083 active cells in the model.

The computer model solves for hydraulic head for each cell and computes ground-water fluxes between cells and within individual cells between the elastic and inelastic storage components. These values are then used to calculate the amount of compaction, if any, for each cell. The total amount of compaction in all three layers is the computed land subsidence at that location.

Differences between measured and simulated compaction and water levels are minimized through a history-matching process. Model parameters are iteratively adjusted to find the best match between simulated and measured values. Once the set of possible model parameters is constrained by the history match, the model may be used cautiously to predict future land subsidence.

In this example, 18 sites were used to match water levels and 10 sites were used to match compaction; one of each is shown.

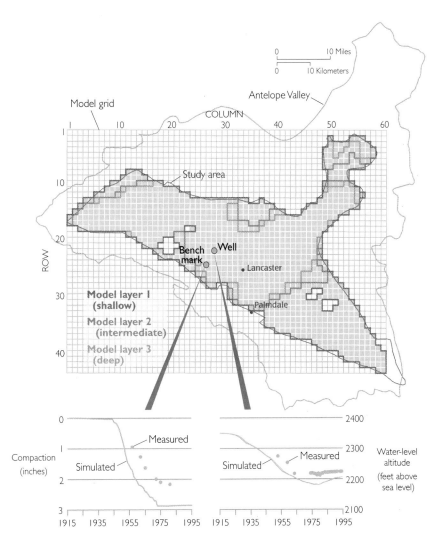

Simulation models are useful analytical tools

Since the advent of high-speed digital computers, scientists have had the ability to numerically simulate the flow of ground water and associated aquifer-system compaction in multiple dimensions. In actual practice, ground-water flow may be simulated in one, two, or three dimensions, but compaction is typically simulated as a one-dimensional process (Helm, 1978). Though poroelastic theory developed by Biot (1941) provides a means for coupling ground-water flow and skeletal deformation in three dimensions, scientists commonly invoke the one-dimensional theory of hydrodynamic consolidation developed by Terzaghi (1925), which is described in some detail in the introduction to the section on "Mining Ground Water." Multidimensional flow of ground water is described by a variant of the well-known diffusion equation that also describes conduction of heat and electricity.

Simulation models can aid visualization of complex three-dimensional processes. They are important analytical tools, and can also be used to help devise data-acquisition and water-management strategies. Though simulation models are powerful tools, it is important to recognize their limitations. The common assumption of one-dimensional consolidation is motivated by an obvious truism: most aquifer-system compaction takes place in the vertical dimension. Nevertheless, the widespread occurrence of earth fissures indicates that horizontal deformation can be locally significant. A more general and more important limitation of simulation models is that their solutions are nonunique. The relevant hydrogeologic parameters (permeabilities, compressibilities, and boundary conditions) are never exactly known, which would be required for a unique solution. Nevertheless, simulation models may be used—with caution—in a predictive mode, and there are formal procedures for dealing with parameter uncertainties.

InSAR images offer new insights

In the Antelope Valley, Mojave Desert, California, a radar interferogram for the period October 20, 1993 to December 22, 1995 revealed up to 2 inches of subsidence in areas previously affected by as much as 6 feet of subsidence since 1930 (Galloway and others, 1998). The regions of maximum subsidence detected during the 26-month period correlated well with declining ground-water levels. In another part of Antelope Valley formerly affected by ground-water depletion and subsidence, but where ground-water levels recovered throughout the 1990s, about 1 inch of subsidence was detected on the interferogram. This suggests residual (time-delayed) compaction due to the presence of thick aquitards. Computer simulations of aquifer-system compaction compared favorably with the subsidence detected by the interferogram for the same period. The computer simulation was weakly constrained due to the scarcity of conventional field measurements; these results highlight the potential use of spatially detailed InSAR subsidence measurements to provide better constraints for computer simulations of land subsidence.

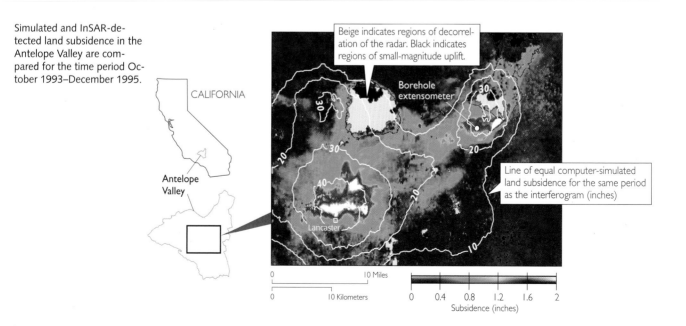

Simulated and InSAR-detected land subsidence in the Antelope Valley are compared for the time period October 1993–December 1995.

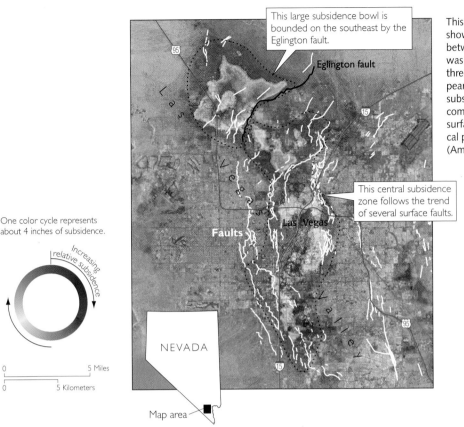

This large subsidence bowl is bounded on the southeast by the Eglington fault.

Eglington fault

This central subsidence zone follows the trend of several surface faults.

Faults

Las Vegas

Las Vegas Valley

NEVADA

Map area

This InSAR-based surface-deformation map shows subsidence in the Las Vegas Valley between April 1992 and December 1997. It was obtained by stacking or summing the three time-sequential interferograms appearing on the cover of this Circular. The subsidence is caused by aquifer-system compaction and controlled in part by the surface faults, which have also been the focal point of earth-fissure formation (Amelung and others, 1999).

One color cycle represents about 4 inches of subsidence.

Increasing relative subsidence

0   5 Miles

0   5 Kilometers

New InSAR-based surface-deformation maps for Las Vegas Valley demonstrate the intimate connection between faults and subsidence. An interferogram for the Las Vegas Valley between April 21, 1992, and December 5, 1997, delineates two main features—a subsidence bowl in the northwest and an elongated subsiding zone in the central part of the valley. The northwest subsidence bowl is nearly circular along its western extent and includes the area of maximum subsidence of nearly 7.5 inches. Its southeastern boundary is aligned along the Eglington scarp, one of several Quaternary faults cutting the valley-floor alluvium. Little subsidence is detected immediately southeast of the fault. Similarly, the central subsidence zone follows the general trends of several mapped faults. The map suggests that the spatial distribution of land subsidence in Las Vegas Valley is controlled by Quaternary faults to a much greater degree than previously suspected. The faults may separate compressible from less-compressible deposits, or they may act as barriers to ground-water flow, impeding the horizontal propagation of fluid-pressure changes and creating ground-water-level differences across the faults.

The potential for renewed subsidence in Santa Clara Valley, California, is a concern for the Santa Clara Valley Water District. One of the District's objectives in managing water resources is to limit ground-water extractions that would cause inelastic (irreversible) compaction of the valley's aquifer system. Seasonal and longer-term elevation changes were measured from successive satellite radar

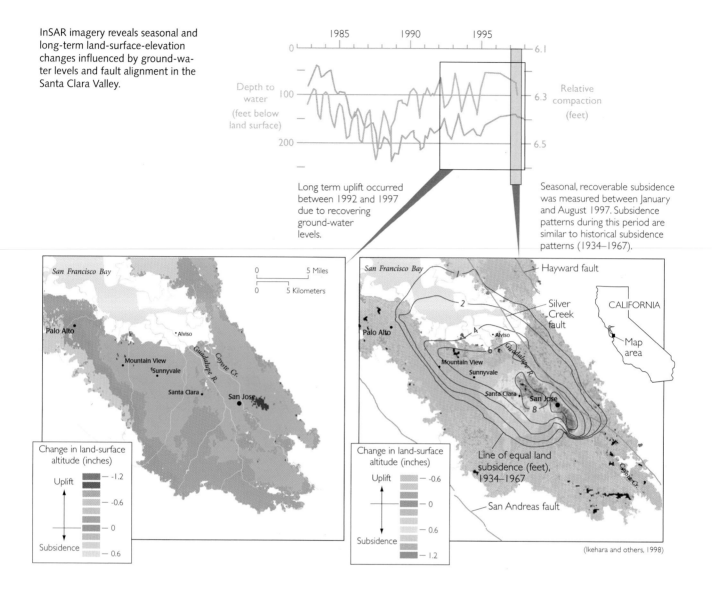

InSAR imagery reveals seasonal and long-term land-surface-elevation changes influenced by ground-water levels and fault alignment in the Santa Clara Valley.

Long term uplift occurred between 1992 and 1997 due to recovering ground-water levels.

Seasonal, recoverable subsidence was measured between January and August 1997. Subsidence patterns during this period are similar to historical subsidence patterns (1934–1967).

(Ikehara and others, 1998)

passes during 1992 to 1997. The longer-term (~5 year) measurement indicates no change for most of the southwestern Santa Clara Valley and land-surface uplift of up to 1.4 inches in the northern and eastern parts of the valley. This uplift is correlated to the recovery of ground-water levels that has been occurring for several years as a result of reduced pumpage and increased recharge. In contrast, the seasonal (6–8 month) interferograms reveal a large region in San Jose undergoing seasonal elastic deformation related to ground-water-level fluctuations. The eastern extent of this deformation appears to be truncated by a Quaternary fault, the Silver Creek Fault, several miles west and roughly parallel to the tectonically active Hayward Fault. The InSAR maps are generally consistent with compaction measured in borehole extensometers.

MITIGATION MEASURES

When development of natural resources causes subsidence, governments sometimes exercise their power either to prohibit the re-

General approaches to mitigation of subsidence will rarely apply to all types of subsidence.

source development or to control it in ways that minimize damage. This may be done through regulation. With adequate monitoring programs and institutional mechanisms in place, optimal benefits may be achieved for both subsidence mitigation and resource development. The Panel on Land Subsidence (NRC, 1991) found that more research is needed in this area of optimal resource allocation and adaptive approaches to land- and water-use management.

In order to wisely and conjunctively manage our land and water resources we first need to define the relevant interacting processes. In the case of land subsidence and ground-water resources, this means understanding the hydrogeologic framework of the resource as well as the demands or stresses that we place on it. It also means identifying a desired state of the land and water resources—a set of goals and objectives that describe some desirable outcomes. These goals and objectives may require guidelines for decisionmaking (policy) to modify usage of the resources in order to attain the desired state. The selection and management of these policies can be based on measures of the condition of the hydrogeologic system.

Land-subsidence and water-management problems are linked

In a typical basin, ground water is in part a renewable resource; a certain amount may be extracted without seriously depleting the amount of water stored. This is the concept of the "safe yield" of a basin. In subsidence-prone alluvial aquifer systems, unless we wish to mine a significant volume of water in storage in the fine-grained sediments, the volume of water withdrawn cannot greatly exceed the natural and artificial replenishment. It may be necessary to maintain ground-water levels above critical thresholds in subsidence-prone areas within the basin in order to avoid incurring new or additional subsidence. Another important consideration is climatic variability, which affects the amount of water available for natural and artificial replenishment. This restricted concept of "safe yield" addresses only the volume of extracted water with respect to a dynamic equilibrium between the water recharging and discharging a basin. Beyond this, to conserve an aquifer system from a water-quality perspective, it may be necessary to maintain certain minimal flow-through rates.

Because aquifer systems have the capacity to store water, the amount of natural outflow from a basin may not be equal to the amount replenished in the same year. Thus the "optimal yield" of a ground-water basin is not necessarily a constant value. It may vary from year-to-year depending upon the state of the aquifer system and the availability of alternative local and imported supplies. The concept of optimal yield incorporates the dynamic nature of the ground-water basin and the adaptability of the management system (Bear, 1979). However, over the long term, the "annual safe yield" of a basin would be roughly equivalent to its average replenishment.

Adaptability has emerged as a conscious element of institutional design in basin-management programs (Blomquist, 1992). Managing basins according to optimal yield assumptions has allowed water users to respond to changed conditions of water supply, including severe drought and relative abundance. Ground-water management plans typically address both demand and supply by adjusting the demands placed on the water-supply system through conservation and water-rationing programs, by adjusting the supply through conjunctive use of ground water and surface water, and by augmenting the supply through aquifer storage and recovery programs. Adaptable management alternatives contribute to the stability and sustainability of land and water resources in many basins.

In basins susceptible to detrimental effects related to the lowering of ground-water levels, such as the three types of subsidence presented in this Circular, land and water resources are linked. For alluvial ground-water basins subject to aquifer-system compaction, threshold values of aquifer-system stress define the boundary between nonpermanent (recoverable) and permanent (nonrecoverable) compaction and loss of land-surface elevation. For regions affected by the dissolution and collapse of soluble rocks, the threshold stress values are more ambiguous but nevertheless real and somewhat manageable. For oxidation of organic soils, the threshold for detrimental effects is very nearly defined by the position of the water table. In each case, management of the land-subsidence problem is inextricably linked to other facets of water-resource management.

Socioeconomic risks versus benefits

Ground-water basins have value not only as perennial sources of water supply, but also as reservoirs for cyclical recharge and discharge. While augmenting base water-supply needs met from a variety of water sources, ground-water basins may provide water at peak-demand periods to modulate the variability inherent in surface-water supplies. The conjunctive surface- and ground-water-management programs in some southern California basins make more efficient use of basin storage capacity than the fixed-yield management programs of other, nonconjunctively managed basins. Storing water underground in wet years for use in dry years, and encouraging water users to take more surface and imported water when it is plentiful and to pump more ground water when it is not, capitalizes on one of the strengths of the ground-water resource. Restricting pumping to the same amount each year regardless of basin conditions does not. In some cases, the most valuable use of ground-water basins is to lessen the immediate shock of short-term variability of water supply (Blomquist, 1992). In subsidence-prone basins, the need to maintain minimum water levels for subsidence control may place a significant constraint on conjunctive-use schemes.

GLOSSARY

Land Subsidence in the United States

These definitions are based on the American Geological Institute's **Glossary of Geology** (4th edition) and **Glossary of Hydrology**, and USGS Water Supply Paper 2025, "Glossary of selected terms useful in studies of the mechanics of aquifer systems and land subsidence due to fluid withdrawal" (Poland, and others, 1971).

Aquifer
A saturated, permeable, geologic unit that can transmit significant quantities of ground water under ordinary hydraulic gradients and is permeable enough to yield economic quantities of water to wells.

Aquifer, Artesian
See *Aquifer, Confined, and Artesian.*

Aquifer, Confined
An artesian aquifer that is confined between two aquitards.

Aquifer, Unconfined
A water-table aquifer in which the water table forms the upper boundary.

Aquifer System
A heterogeneous body of interbedded permeable and poorly permeable geologic units that function as a water-yielding hydraulic unit at a regional scale. The aquifer system may comprise one or more aquifers within which aquitards are interspersed. Confining units may separate the aquifers and impede the vertical exchange of ground water between aquifers within the aquifer system.

Aquitard
A saturated, but poorly permeable, geologic unit that impedes ground-water movement and does not yield water freely to wells, but which may transmit appreciable water to and from adjacent aquifers and, where sufficiently thick, may constitute an important ground-water storage unit. Areally extensive aquitards may function regionally as confining units within aquifer systems. See also *Confining Unit.*

Artesian
An adjective referring to confined aquifers. Sometimes the term artesian is used to denote a portion of a confined aquifer where the altitudes of the potentiometric surface are above land surface (flowing wells and artesian wells are synonymous in this usage). But more generally the term indicates that the altitudes of the potentiometric surface are above the altitude of the base of the confining unit (artesian wells and flowing wells are not synonymous in this case). See *Aquifer, Confined.*

Blue hole
A subsurface void, usually a solution sinkhole, developed in carbonate rocks that are open to the Earth's surface and contains tidally influenced waters of fresh, marine, or mixed chemistry.

Cave
A natural underground open space or series of open spaces and passages large enough for a person to enter, generally with a connection to the surface; often formed by solution of limestone.

Cavern	A cave, with the implication of a large size.
Cenote	Steep-walled natural well that extends below the water table; generally caused by collapse of a cave roof; term reserved for features found in the Yucatan Peninsula of Mexico.
Confining Unit	A saturated, relatively low-permeability geologic unit that is areally extensive and serves to confine an adjacent artesian aquifer or aquifers. Leaky confining units may transmit appreciable water to and from adjacent aquifers. See also *Aquitard*.
Compaction	In this Circular, compaction is used in its geologic sense and refers to the inelastic compression of the aquifer system. Compaction of the aquifer system reflects the rearrangement of the mineral grain pore structure and largely nonrecoverable reduction of the porosity under stresses greater than the preconsolidation stress. Compaction, as used here, is synonymous with the term "virgin consolidation" used by soils engineers. The term refers to both the process and the measured change in thickness. As a practical matter, a very small amount (1 to 5 percent) of the compaction is recoverable as a slight elastic rebound of the compacted material if stresses are reduced.
Compaction, Residual	Compaction that would ultimately occur if a given increase in applied stress were maintained until steady-state pore pressures were achieved. Residual compaction may also be defined as the difference between (1) the amount of compaction that will occur ultimately for a given increase in applied stress, and (2) that which has occurred at a specified time.
Compression	In this Circular, compression refers to the decrease in thickness of sediments, as a result of increase in vertical compressive stress. Compression may be elastic (fully recoverable) or inelastic (nonrecoverable).
Consolidation	In soil mechanics, consolidation is the adjustment of a saturated soil in response to increased load, involving the squeezing of water from the pores and a decrease in void ratio or porosity of the soil. In this Circular, the geologic term "compaction" is used in preference to consolidation.
Datum	See *Geodetic Datum*.
Ellipsoid, Earth	A mathematically determined three-dimensional surface obtained by rotating an ellipse about its semi-minor axis. In the case of the Earth, the ellipsoid is the modeled shape of its surface, which is relatively flattened in the polar axis.
Ellipsoid, Height	The distance of a point above the ellipsoid measured perpendicular to the surface of the ellipsoid.
Exfoliation	The process by which concentric scales, plates, or shells of rock, from less than a centimeter to several meters in thickness, are stripped from the bare surface of a large rock mass. See *spall*.
Geodetic Datum	A set of constants specifying the coordinate system used for geodetic control, for example, for calculating the coordinates of points on the Earth.

Geoid, Earth	The sea-level equipotential surface or figure of the Earth. If the Earth were completely covered by a shallow sea, the surface of this sea would conform to the geoid shaped by the hydrodynamic equilibrium of the water subject to gravitational and rotational forces. Mountains and valleys are departures from this reference geoid.
Head, Hydraulic	A measure of the potential for fluid flow. The height of the free surface of a body of water above a given subsurface point.
Hydraulic Conductivity	A measure of the medium's capacity to transmit a particular fluid. The volume of water at the existing kinematic viscosity that will move in a porous medium in unit time under a unit hydraulic gradient through a unit area. In contrast to permeability, it is a function of the properties of the liquid as well as the porous medium.
Hydrocompaction	The process of volume decrease and density increase that occurs when certain moisture-deficient deposits compact as they are wetted for the first time since burial. The vertical downward movement of the land surface that results from this process has also been termed "shallow subsidence" and "near-surface subsidence."
Karst	A type of topography that is formed on limestone, dolomite, gypsum and other rocks, primarily by dissolution, and that is characterized by sinkholes, caves, and subterranean drainage.
Karstification	Action by water, mainly chemical but also mechanical, that produces features of a karst topography.
Karst, Mantled	A terrane of karst features, usually subdued, and covered by soil or a thin alluvium.
Load	We refer to *Load* as synonymous with *Stress*.
Overdraft	Any withdrawal of ground water in excess of the *Safe Yield*.
Paleokarst	A karstified area that has been buried by later deposition of sediments.
Permeability	The capacity of a porous rock, sediment, or soil for transmitting a fluid. Unlike hydraulic conductivity, it is a function only of the medium.
pH	A measure of the acid/base property of a material sample. The negative logarithm of the hydrogen ion concentration; pH 7 is neutral with respect to distilled, deionized water; pH less than 7 is more acidic; pH greater than 7 is more basic.
Piezometric Surface	See *Potentiometric Surface*.
Plutonic	A loosely defined term with a number of current usages. We use it to describe igneous rock bodies that crystallized at great depth or, more generally, any intrusive igneous rock.
Porosity	The percentage of the soil or rock volume that is occupied by pore space, void of material. The porosity is defined by the ratio of void space to the total volume of a specimen.
Potentiometric Surface	An imaginary surface representing the total head of ground water and defined by the level to which the water will rise in a tightly cased well. See *Head, Hydraulic*.

Recharge	The process involved in addition of water to the saturated zone, naturally by precipitation or runoff, or artificially by spreading or injection.
Sinkhole	A depression in a karst area. At land surface its shape is generally circular and its size measured in meters to tens of meters; underground it is commonly funnel-shaped and associated with subterranean drainage.
Safe Yield	See *Yield, Safe.*
Specific Storage	The volume of water that an aquifer system releases or takes into storage per unit volume per unit change in head. The specific storage is equivalent to the *Storage Coefficient* divided by the thickness of the aquifer system.
Spall	A chip or fragment removed from a rock surface by weathering; especially by the process of exfoliation. See *exfoliation.*
Spring	Any natural discharge of water from rock or soil onto the land surface or into a surface-water body.
Storage	The capacity of an aquifer, aquitard, or aquifer system to release or accept water into ground-water storage, per unit change in hydraulic head. See *Storage Coefficient* and *Specific Storage.*
Storage Coefficient	The volume of water that an aquifer system releases or takes into storage per unit surface area per unit change in head.
Strain	Relative change in the volume, area or length of a body as a result of *stress*. The change is expressed in terms of the amount of displacement measured in the body divided by its original volume, area, or length, and referred to as either a volume strain, areal strain, or one-dimensional strain, respectively. The unit measure of strain is dimensionless, as its value represents the fractional change from the former size.
Stress	In a solid body, the force (per unit area) acting on any surface within it; also refers to the applied force (per unit area) that creates the internal force. Stress is variously expressed in units of pressure, such as pounds per square inch, kilograms per square meter, or Pascals.
Stress, Applied	The downward stress imposed on a specified horizontal plane within an aquifer system. At any given level in the aquifer system, the applied stress is the force or weight (per unit area) of sediments and moisture above the water table, plus the submerged weight (per unit area), accounting for buoyancy of the saturated sediments overlying the specified plane at that level, plus or minus the net seepage stress generated by flow (upward or downward component) through the specified plane in the aquifer system.
Stress, Effective	Stress (pressure) that is borne by and transmitted through the grain-to-grain contacts of a deposit, and thus affects its porosity and other physical properties. In one-dimensional compression, effective stress is the average grain-to-grain load per unit area in a plane normal to the applied stress. At any given depth, the effective

stress is the weight (per unit area) of sediments and moisture above the water table, plus the submerged weight (per unit area) of sediments between the water table and the specified depth, plus or minus the seepage stress (hydrodynamic drag) produced by downward or upward components, respectively, of water movement through the saturated sediments above the specified depth. Effective stress may also be defined as the difference between the geostatic stress and fluid pressure at a given depth in a saturated deposit, and represents that portion of the applied stress which becomes effective as intergranular stress.

Stress, Geostatic (Lithostatic)   The total weight (per unit area) of sediments and water above some plane of reference. Geostatic stress normal to any horizontal plane of reference in a saturated deposit may also be defined as the sum of the effective stress and the fluid pressure at that depth.

Stress, Preconsolidation   The maximum antecedent effective stress to which a deposit has been subjected and which it can withstand without undergoing additional permanent deformation. Stress changes in the range less than the preconsolidation stress produce elastic deformations of small magnitude. In fine-grained materials, stress increases beyond the preconsolidation stress produce much larger deformations that are principally inelastic (nonrecoverable). Synonymous with "virgin stress."

Stress, Seepage   Force (per unit area) transferred from the water to the medium by viscous friction when water flows through a porous medium. The force transferred to the medium is equal to the loss of hydraulic head and is termed the seepage force exerted in the direction of flow.

Subsidence   Sinking or settlement of the land surface, due to any of several processes. As commonly used, the term relates to the vertical downward movement of natural surfaces although small-scale horizontal components may be present. The term does not include landslides, which have large-scale horizontal displacements, or settlements of artificial fills.

Subsidence, Near-Surface   See *Hydrocompaction.*

Subsidence, Shallow   See *Hydrocompaction.*

Transmissivity   The rate at which water at the prevailing kinematic viscosity is transmitted through a unit width of aquifer under a unit hydraulic gradient. See also *Hydraulic Conductivity.*

Vug   A small cavity or chamber in rock that may be lined with crystals.

Water Table   The surface of a body of unconfined ground water at which the pressure is equal to atmospheric pressure.

Yield, Operational   See *Yield, Optimal.*

Yield, Optimal   An optimal amount of ground water, by virtue of its use, that should be withdrawn from an aquifer system or ground-water basin each year. It is a dynamic quantity that must be determined from a set of alternative ground-water management decisions subject to goals, objectives, and constraints of the management plan.

Yield, Perennial   The amount of usable water from an aquifer that can be economically consumed each year for an indefinite period of time. It is a specified amount that is commonly specified equal to the mean annual recharge to the aquifer system, which thereby limits the amount of ground water that can be pumped for beneficial use.

Yield, Safe   The amount of ground water that can be safely withdrawn from a ground-water basin annually, without producing an undesirable result. Undesirable results include but are not limited to depletion of ground-water storage, the intrusion of water of undesirable quality, the contraventions of existing water rights, the deterioration of the economic advantages of pumping (such as excessively lowered water levels and the attendant increased pumping lifts and associated energy costs), excessive depletion of streamflow by induced infiltration, and land subsidence.

REFERENCES

Land Subsidence in the United States

Introduction

Ege, J.R., 1984, Mechanisms of surface subsidence resulting from solution extraction of salt, *in* Holzer, T.L., ed., Man-induced land subsidence: Geological Society of America Reviews in Engineering Geology, v. 6, p. 203–221.

Lucas, R.E., 1982, Organic soils (Histosols)—Formation, distribution, physical and chemical properties and management for crop production: Michigan State University Farm Science Research Report 435, 77 p.

National Research Council, 1991, Mitigating losses from land subsidence in the United States: Washington, D. C., National Academy Press, 58 p.

Stephens, J.C., Allen, L.H., Jr., and Chen, Ellen, 1984, Organic soil subsidence, *in* Holzer, T.L., ed., Man-induced land subsidence: Geological Society of America Reviews in Engineering Geology, v. 6, p. 107–122.

White, W.B., Culver, D.C., Herman, J.S., Kane, T.C., and Mylroie, J.E., 1995, Karst lands: American Scientist, v. 83, p. 450–459.

PART I—Mining Ground Water

INTRODUCTION

Clawges, R. M., and Price, C. V., 1999, Digital data sets describing principal aquifers, surficial geology, and ground-water regions of the conterminous United States: U.S. Geological Survey Open-File Report 99-77 [accessed Sept. 17, 1999 at URL http://water.usgs.gov/pubs/ofr/ofr99-77].Freeze, R.A., and Cherry, J.A., 1979, Groundwater: Englewood Cliffs, N. J. Prentice-Hall, 604 p.

Green, J.H., 1964, Compaction of the aquifer system and land subsidence in the Santa Clara Valley, California: U.S. Geological Survey Water-Supply Paper 1779-T, 11 p.

Helm, D.C., 1975, One-dimensional simulation of aquifer system compaction near Pixley, Calif., part 1. Constant parameters: Water Resources Research, v. 11, p. 465–478.

Heywood, C.E., 1997, Piezometric-extensometric estimations of specific storage in the Albuquerque Basin, New Mexico: *in* Prince, K.R., and Leake, S.A., eds., U.S. Geological Survey Open-File Report 97–47, p. 21–26.

Holzer, T.L., 1998, History of the aquitard-drainage model in land subsidence case studies and current research, *in* Borchers, J.W., ed., Land subsidence case studies and current research: Proceedings of the Dr. Joseph F. Poland symposium on land subsidence, Association of Engineering Geologists Special Publication No. 8, p. 7–12.

Ireland, R.L., Poland, J.F., and Riley, F.S., 1984, Land subsidence in the San Joaquin Valley, California as of 1980: U.S. Geological Survey Professional Paper 437-I, 93 p.

Miller, R.E., 1961, Compaction of an aquifer system computed from consolidation tests and decline in artesian head: U.S. Geological Survey Professional Paper 424-B, p. B54–B58.

Poland, J.F., 1960, Land subsidence in the San Joaquin Valley and its effect on estimates of ground-water resources: International Association of Scientific Hydrology, IASH Publication 52, p. 324–335.

Poland, J.F., and Green, J.H., 1962, Subsidence in the Santa Clara Valley, California—a progress report: U.S. Geological Survey Water-Supply Paper 1619-C, 16 p.

Poland, J.F., Lofgren, B.E., Ireland, R.L., and Pugh, R.G., 1975, Land subsidence in the San Joaquin Valley, California as of 1972: U.S. Geological Survey Professional Paper 437-H, 78 p.

Poland, J.F., ed., 1984, Guidebook to studies of land subsidence due to ground-water withdrawal: United Nations Educational, Scientific and Cultural Organization, Paris, Studies and reports in hydrology 40, 305 p.

Poland, J.F., and Ireland, R.L., 1988, Land subsidence in the Santa Clara Valley, California, as of 1982: U.S. Geological Survey Professional Paper 497-F, 61 p.

Riley, F.S., 1969, Analysis of borehole extensometer data from central California: International Association of Scientific Hydrology Publication 89, p. 423–431.

Riley, F.S., 1998, Mechanics of aquifer systems—The scientific legacy of Joseph F. Poland, in Borchers, J.W., ed., Land subsidence case studies and current research: Proceedings of the Dr. Joseph F. Poland symposium on land subsidence, Association of Engineering Geologists Special Publication No. 8, p. 13–27.

Terzaghi, K., 1925, Principles of soil mechanics, IV—Settlement and consolidation of clay: Engineering News-Record, 95(3), p. 874–878.

Tolman, C.F., and Poland, J.F., 1940, Ground-water infiltration, and ground-surface recession in Santa Clara Valley, Santa Clara County, California: Transactions American Geophysical Union, v. 21, p. 23–34.

SANTA CLARA VALLEY, CALIFORNIA

California History Center, 1981, Water in the Santa Clara Valley—A history: De Anza College California History Center Local History Studies v. 27, 155 p.

Fowler, L.C., 1981, Economic consequences of land surface subsidence: Journal of the Irrigation and Drainage Division, American Society of Civil Engineers, v. 107, p. 151–159.

Poland, J.F., 1977, Land subsidence stopped by artesian head recovery, Santa Clara Valley, California: International Association of Hydrological Sciences Publication 121, p. 124–132.

Poland, J.F., and Ireland, R.L., 1988 , Land subsidence in the Santa Clara Valley, California, as of 1982: U.S. Geological Survey Professional Paper 497-F, 61 p.

Reichard, E.G., and Bredehoeft, J.D., 1984, An engineering economic analysis of a program for artificial groundwater recharge: Water Resources Bulletin, v. 20, p. 929–939.

Tolman, C.F., and Poland, J.F. 1940, Ground-water infiltration, and ground-surface recession in Santa Clara Valley, Santa Clara County, California: Transactions American Geophysical Union, v. 21, p. 23–34.

SAN JOAQUIN VALLEY, CALIFORNIA

Cone, Tracy, 1997, The vanishing valley: San Jose Mercury News West Magazine, June 29, p. 9–15.

EDAW-ESA, 1978, Environmental and economic effects of subsidence: Lawrence Berkeley Laboratory Geothermal Subsidence Research Program Final Report—Category IV, Project 1, [variously paged].

Ingerson, I. M., 1941, The hydrology of the southern San Joaquin Valley, California, and its relation to imported water supplies: Transactions American Geophysical Union, v. 22, p. 20–45.

Ireland, R.L., Poland, J.F., and Riley, F.S., 1984, Land subsidence in the San Joaquin Valley, California as of 1980: U.S. Geological Survey Professional Paper 437-I, 93 p.

Manning, J.C., 1967, Report on the ground-water hydrology in the southern San Joaquin Valley: American Water Works Association Journal, v. 59, p. 1,513–1,526.

Mendenhall, W.C., Dole, R.B., and Stabler, Herman, 1916, Ground water in San Joaquin Valley, California: U.S. Geological Survey Water-Supply Paper 398, 310 p.

Nady, Paul, and Larragueta, L.L., 1983, Development of irrigation in the Central Valley of California: U.S. Geological Survey Hydrologic Investigations Atlas HA-649, scale 1:500,000, 2 sheets.

Poland, J.F., Lofgren, B.E., Ireland, R.L., and Pugh, R.G., 1975, Land subsidence in the San Joaquin Valley, California, as of 1972: U.S. Geological Survey Professional Paper 437-H, 78 p.

Swanson, A.A., 1998, Land subsidence in the San Joaquin Valley, updated to 1995, *in* Borchers, J.W., ed., Land subsidence case studies and current research: Proceedings of the Dr. Joseph F. Poland symposium on land subsidence, Association of Engineering Geologists Special Publication No. 8, p. 75–79.

Williamson, A.K., Prudic, D.E., and Swain, L.A., 1989, Ground-water flow in the Central Valley, California: U.S. Geological Survey Professional Paper 1401-D, 127 p.

HOUSTON-GALVESTON, TEXAS

Gabrysch, R.K., 1983, The impact of land-surface subsidence, *in* Impact of Science on Society, Managing our fresh-water resources: United Nations Educational, Scientific and Cultural Organization No. 1, p. 117–123.

Galveston Bay National Estuary Program, 1995, The Galveston Bay plan—The comprehensive land management plan for the Galveston Bay ecosystem, Oct. 18, 1994: Galveston Bay National Estuary Program Publication GBNEP-49, 457 p.

Holzer, T.L., 1984, Ground failure induced by ground water withdrawal from unconsolidated sediment, *in* Holzer, T.L., ed., Man-induced land subsidence: Geological Society of America Reviews in Economic Geology, v. 6, p. 67–105.

Holzer, T.L., and Gabrysch, R.K., 1987, Effect of water-level recoveries on fault creep, Houston, Texas: Ground Water, v. 25, p. 392–397.

Holzschuh, J.C., 1991, Land subsidence in Houston, Texas U.S.A.: Field-trip guidebook for the fourth international symposium on land subsidence, May 12–17, 1991, Houston, Tex., 22 p.

Kasmarek, M.C., Coplin, L.S., and Santos, H.X., 1997, Water-level altitudes 1997, water-level changes 1977–97 and 1996–97, and compaction 1973–96 in the Chicot and Evangeline Aquifers, Houston-Galveston Region, Texas: U.S. Geological Survey Open-File Report 97-181, 8 sheets, scale 1:100,000.

McGowen, J.M., Garner, L.E., and Wilkinson, B.M., 1977, The Gulf shoreline of Texas—processes, characteristics, and factors in use: University of Texas, Bureau of Economic Geology, Geological Circular 75-6, 43 p.

Jones, L.L., 1976, External costs of surface subsidence— Upper Galveston Bay, Texas: International Symposium on Land Subsidence, 2nd, Anaheim, Calif., December 1976, [Proceedings] International Association of Hydrological Sciences Publication 121, p. 617–627.

Paine, J.G., 1993, Subsidence of the Texas coast—Inferences from historical and late Pleistocene sea levels: Tectonophysics, v. 222, p. 445–458.

Paine, J.G., and Morton, R.A., 1986, Historical shoreline changes in Trinity, Galveston, West, and East Bays, Texas Gulf Coast: University of Texas Bureau of Economic Geology Circular 86-3, 58 p.

Pratt, W.E., and Johnson, D.W., 1926, Local subsidence of the Goose Creek oil field: Journal of Geology, v. 34, p. 577–590.

Titus, J.G., and Narayanan, V.K., 1995, The probability of sea level rise: U.S. Environmental Protection Agency, EPA 230-R-95-008.

White, W.A., Tremblay, T.A., Wermund, E.G., Jr., and Handley, L.R., 1993, Trends and status of wetland and aquatic habitats in the Galveston Bay system, Texas: Galveston Bay National Estuary Program Publication GBNEP-31, 225 p.

LAS VEGAS, NEVADA

Acevedo, William, Gaydos, Leonard, Tilley, Janet, Mladinich, Carol, Buchanan, Janis, Blauer, Steve, Kruger, Kelley, and Schubert, Jamie, 1997, Urban land use change in the Las Vegas Valley: U.S. Geological Survey, accessed July 27, 1999, http://geochange.er.usgs.gov/sw/changes/anthropogenic/population/las_vegas.

Bell, J.W., 1981a, Subsidence in Las Vegas Valley: Nevada Bureau of Mines and Geology Bulletin 95, 83 p., 1 plate, scale 1:62,500.

Bell, J.W., 1981b, Results of leveling across fault scarps in Las Vegas Valley, Nevada, April 1978–June 1981: Nevada Bureau of Mines and Geology Open-File Report 81-5, 7 p.

Bell, J.W., and Helm, D.C., 1998, Ground cracks on Quaternary faults in Nevada—Hydraulic and tectonic, in Borchers, J.W., ed., Land subsidence case studies and current research: Proceedings of the Dr. Joseph F. Poland symposium on land subsidence, Association of Engineering Geologists Special Publication No. 8, p. 165–173.

Bell, J.W., and Price, J.G., 1991, Subsidence in Las Vegas Valley, 1980–91—Final project report: Nevada Bureau of Mines and Geology, Open-File Report 93-4, 10 sect., 9 plates, scale 1:62,500.

Bernholtz, A., Brothers, K., and Katzer, T., 1994, Artificial ground-water recharge in Las Vegas Valley, Clark County, Nevada— Storing today, treating tomorrow?: International Symposium on Artificial Recharge of Ground Water, 2nd, Orlando, Fla., July 1994, [Proceedings], p. 548–557.

Burbey, T.J., 1995, Pumpage and water-level change in the principal aquifer of Las Vegas Valley, 1980–90—Nevada Division of Water Resources Information Report 34, 224 p.

Carpenter, Everett, 1915, Ground water in southeastern Nevada: U.S. Geological Survey Water-Supply Paper 365, 86 p.

Coache, Robert, 1996, Las Vegas Valley water usage report, Clark County, Nevada, 1996: Nevada Division of Water Resources Report, [50+] p.

Dettinger, M.D., 1989, Reconnaissance estimates of natural recharge to desert basins in Nevada, U.S.A., by using chloride balance calculations: Journal of Hydrology, v. 106, p. 55–78.

Domenico, P.A., Stephenson, D.A., and Maxey, G.B., 1964, Ground water in Las Vegas Valley: Nevada Department of Conservation and Natural Resources Division of Water Resources Bulletin 29, 53 p.

Hafen, L.R., and Hafen, A.W., 1954, Old Spanish Trail, Santa Fe to Los Angeles—With extracts from contemporary records and including diaries of Antonio Armijo and Orville Pratt: University of Nebraska Press, 375 p.

Harrill, James R., 1976, Pumping and ground-water storage depletion in Las Vegas Valley, Nevada, 1955–74: Nevada Department of Conservation and Natural Resources, Division of Water Resources Bulletin No. 44, 70 p.

Holzer, T.L., 1979, Leveling data—Eglington fault scarp, Las Vegas Valley, Nevada: U.S. Geological Survey Open-File Report 79-950, 7 p.

— 1984, Ground failure induced by ground-water withdrawal from unconsolidated sediment, in Holzer, T.L., ed., Man-induced land subsidence: Geological Society of America Reviews in Engineering Geology, v. 6, 221 p.

Houghton, J. G., Sakamoto, C. M., and Gifford, R.O., 1975, Nevada's weather and climate: Nevada Bureau of Mines and Geology Special Publication 2, 78 p.

Jones, F. L., and Cahlan, J. F. 1975, Water, a history of Las Vegas—Volume I: Las Vegas Valley Water District, 171 p.

Livingston, Penn, 1941, Underground leakage from artesian wells in the Las Vegas area, Nevada: U.S. Geological Survey Water-Supply Paper 849-D, p. 147–173.

Malmberg, G. T., 1965, Available water supply of the Las Vegas ground-water basin, Nevada: U.S. Geological Survey Water-Supply Paper 1780, 116 p., 13 plates.

Maxey, G.B., and Jameson, C.H., 1948, Geology and water resources of Las Vegas, Pahrump, and Indian Springs Valleys, Clark and Nye Counties, Nevada: Nevada State Engineer Water Resources Bulletin 5, 121 p.

Mendenhall, W.C., 1909, Some desert watering places in southeastern California and southwestern Nevada: U.S. Geological Survey Water-Supply Paper 224, 98 p.

Mifflin, M.D., and Wheat, M.M., 1979, Pluvial lakes and estimated pluvial climates of Nevada: Nevada Bureau of Mines and Geology Bulletin 94, 57 p., 1 plate.

Mindling, A.L., 1971, A summary of data relating to land subsidence in Las Vegas Valley: Center for Water Resources Research, Desert Research Institute, University of Nevada, Reno; 55 p.

Morgan, D.S., and Dettinger, M.D., 1996, Ground-water conditions in Las Vegas Valley, Clark County, Nevada, part 2, Hydrogeology and simulation of ground-water flow: U.S. Geological Survey Water-Supply Paper 2320-B, 124 p., 2 plates.

Nevada Department of Conservation and Natural Resources, Division of Water Resources and Water Planning, 1992, Hydrographic basin summaries, 1990–1992, [variously paged].

Plume, R. W., 1989, Ground-water conditions in Las Vegas Valley, Clark County, Nevada, part I, Hydrogeologic framework: U.S. Geological Survey Water-Supply Paper 2320-A, 15 p.

Quade, J., Mifflin, M.D., Pratt, W.L., McCoy, W., and Burckle, L., 1995, Fossil spring deposits in the southern Great Basin and their implications for change in water-table levels near Yucca Mountain, Nevada, during Quaternary time: Geological Society of America Bulletin, v. 107, p. 213–230.

Riley, F.S., 1969, Analysis of borehole extensometer data from central California: International Association of Scientific Hydrology Publication 89, p. 423–431.

U.S. Department of Commerce, 1997, Las Vegas metro area leads nation in population growth, census bureau reports: Bureau of the Census, accessed July 27, 1999, at URL http://www.census.gov/ftp/pub/Press-Release/cb97-212.html

Water Resources Management Incorporated, 1992, WRMI process—Water supply planning for the Las Vegas region, Columbia, Md., [25+] p.

SOUTH-CENTRAL ARIZONA

Anderson, S.R., 1988, Potential for aquifer compaction, land subsidence, and earth fissures in the Tucson Basin, Pima County, Arizona: U.S. Geological Survey Hydrologic Investigations Atlas 713, 3 sheets, scale 1:250,000.

Anderson, S.R., 1989, Potential for aquifer compaction, land subsidence, and earth fissures in Avra Valley, Pima and Pinal Counties, Arizona: U.S. Geological Survey Hydrologic Investigations Atlas 718, 3 sheets, scale 1:250,000.

Anderson, T.W., Freethey, G.W., and Tucci, P., 1992, Geohydrology and water resources of alluvial basins in south-central Arizona and parts of adjacent states: U.S. Geological Survey Professional Paper 1406-B, 67 p., 3 plates, scale 1:1,000,000.

Anning, D.W., and Duet, N.R., 1994, Summary of ground-water conditions in Arizona, 1987–90: U.S. Geological Survey Open-File Report 94-476, 2 sheets.

Arizona Department of Water Resources, 1999, Arizona water information, statewide overview, supply and demand (1994): accessed July 27, 1999 at URL http://www.adwr.state.az.us/AZWaterInfo/statewide/supplyde.html.

Carpenter, M.C., 1993, Earth-fissure movements associated with fluctuations in ground-water levels near the Picacho Mountains, south-central Arizona, 1980–84: U.S. Geological Survey Professional Paper 497-H, 49 p.

Carpenter, M.C., and Bradley, M.D., 1986, Legal perspectives on subsidence caused by ground-water withdrawal in Texas, California, and Arizona, U.S.A.: International Symposium on Land Subsidence, 3rd, Venice, 1984, [Proceedings, Johnson, A.I., Carbognin Laura, and Ubertini, L., eds.], International Association of Scientific Hydrology Publication 151, p. 817–828.

City of Tucson Water Department, 1995, Annual static water level basic data report, Tucson Basin and Avra Valley, Pima County, Arizona, 1995: City of Tucson Water Planning and Engineering Division, 140 p.

Eaton, G.P., Peterson, D.L., and Schumann, H.H., 1972, Geophysical, geohydrological, and geochemical reconnaissance of the Luke salt body, Central Arizona: U.S. Geological Survey Professional Paper 753, 28 p.

Feth, J.H., 1951, Structural reconnaissance of the Red Rock quadrangle, Arizona: U.S. Geological Survey Open-File Report, 30 p.

Holzer, T.L., 1980, Reconnaissance maps of earth fissures and land subsidence, Bowie and Willcox areas, Arizona: U.S. Geological Survey Miscellaneous Field Studies Map MF-1156, 2 sheets, scale 1:24,000.

Holzer, T.L., 1984, Ground failure induced by ground-water withdrawal from unconsolidated sediment, in Holzer, T.L., ed., Man-induced land subsidence: Geological Society of America Reviews in Engineering Geology, v. 6, p. 67–105.

Holzer, T.L., Davis, S.N., and Lofgren, B.E., 1979, Faulting caused by groundwater extraction in south-central Arizona: Journal of Geophysical Research, v. 84, p. 603-612.

Johnson, N.M., 1980, The relation between ephemeral stream regime and earth fissuring in south-central Arizona: Tucson, Ariz., University of Arizona, M.S. thesis, 158 p.

Laney, R.L., Raymond, R.H., and Winikka, C.C., 1978, Maps showing water-level declines, land subsidence, and earth fissures in south-central Arizona: U.S. Geological Survey Water-Resources Investigations Report 78-83, 2 sheets, scale 1:125,000.

Leonard, R.J., 1929, An earth fissure in southern Arizona: Journal of Geology, v. 37, p. 765-774.

Peterson, D.E., 1962, Earth fissuring in the Picacho area, Pinal County, Arizona: Tucson, Ariz., University of Arizona, M.S. thesis, 35 p.

Robinson, G.M., and Peterson, D.E., 1962, Notes on earth fissures in southern Arizona: U.S. Geological Survey Circular 466, 7 p.

Schumann, H.H., 1995, Land subsidence and earth fissure hazards near Luke Air Force Base, Arizona: in Prince, K.R., Galloway, D.L., and Leake, S.A., eds., U.S. Geological Survey subsidence interest group conference, Edwards Air Force Base, Antelope Valley, California, November 18–19, 1992—abstracts and summary: U.S. Geological Open-File Report 94-532, p. 18-21.

Schumann, H.H., and Cripe, L.S., 1986, Land subsidence and earth fissures caused by groundwater depletion in southern Arizona, U.S.A.: International Symposium on Land Subsidence, 3rd, Venice, 1984, [Proceedings, Johnson, A.I., Carbognin Laura, and Ubertini, L., eds.], International Association of Scientific Hydrology Publication 151, p. 841–851.

Schumann, H.H., and Genauldi, R.B., 1986, Land subsidence, earth fissures, and water-level change in southern Arizona: Arizona Bureau of Geology and Mineral Technology Geological Survey Branch Map 23, 1 sheet, scale 1:1,000,000.

Schumann, H.H., and Poland, J.F., [1969–1970], Land subsidence, earth fissures, and groundwater withdrawal in south-central Arizona, U.S.A: International Association of Scientific Hydrology Publication 88, p. 295–302.

Strange, W.E., 1983, Subsidence monitoring for the State of Arizona: National Oceanic and Atmospheric Administration National Geodetic Information Center, Rockville, Md., 80 p.

PART II—Drainage of Organic Soils

INTRODUCTION

Darby, H.C., 1956, The Draining of the Fens (2nd ed.): Oxford, Cambridge University Press, 314 p.

Lucas, R.E., 1982, Organic soils (Histosols)—Formation, distribution, physical and chemical properties and management for crop production: Michigan State University Farm Science Research Report 435, 77 p.

Nieuwenhuis, H.S., and Schokking, F., 1997, Land subsidence in drained peat areas of the Province of Friesland, The Netherlands: Quarterly Journal of Engineering Geology, v. 30, p. 37–48.

Schothorst, C.J., 1977, Subsidence of low moor peat soils in the western Netherlands: Geoderma, v. 17, p. 265–291.

Stephens, J.C., Allen, L.H., Jr., and Chen, Ellen, 1984, Organic soil subsidence in Holzer, T.L., ed., Man-induced land subsidence: Geological Society of America Reviews in Engineering Geology, v. 6, p. 107–122.

Waksman, S.A., and Purvis, E.R., 1932, The influence of moisture upon the rapidity of decomposition of lowmoor peat: Soil Science, v. 34, p. 323–336.

Waksman, S.A., and Stevens, K.R., 1929, Contribution to the chemical composition of peat, part 5. The role of microorganisms in peat formation and decomposition: Soil Science, v. 28, p. 315–340.

Wosten, J.H.M., Ismail, A.B., and van Wijk, A.L.M., 1997, Peat subsidence and its practical implications: A case study in Malaysia: Geoderma, v. 78, p. 25–36.

SACRAMENTO-SAN JOAQUIN DELTA, CALIFORNIA

Atwater, B.F., 1980, Attempts to correlate Late Quaternary climatic records between the San Francisco Bay, the Sacramento-San Joaquin Delta, and the Mokelumne River, California: Dover, Del., University of Delaware, Ph.D. dissertation, 215 p.

California Department of Water Resources, 1993, Sacramento-San Joaquin Delta atlas: Sacramento, State of California Department of Water Resources, 121 p.

California Department of Water Resources, 1995, Delta levees: Sacramento, State of California Department of Water Resources, 19 p.

Delta Protection Commission, 1995, Land use and resource management plan for the primary zone of the Delta: Walnut Grove, Delta Protection Commission, 60 p.

Deverel, S.J., and Rojstaczer, S.A., 1996, Subsidence of agricultural lands in the Sacramento-San Joaquin Delta, California: Role of aqueous and gaseous carbon fluxes: Water Resources Research, v. 32, p. 2,359-2,367.

Dillon, Richard, 1982, Delta Country: Novato, Calif., Presidio Press, 134 p.

Rojstaczer, S.A., and Deverel, S.J., 1993, Time dependence of atmospheric carbon inputs from drainage of organic soils: Geophysical Research Letters, v. 20, p. 1,383–1,386.

Rojstaczer, S.A., Hamon, R.E., Deverel, S.J., and Massey, C.A., 1991, Evaluation of selected data to assess the causes of subsidence in the Sacramento-San Joaquin Delta, California: U.S. Geological Survey Open-File Report 91-193, 16 p.

Tans, P.P., Fung, I.Y., and Takahashi, Y., 1990, Observational constraints on the global atmospheric CO2 budget: Science, v. 247, p. 1,431–1,438.

Thompson, John, 1957, The settlement geography of the Sacramento-San Joaquin Delta, California: Palo Alto, Calif., Stanford University, Ph.D. dissertation, 551 p.

Weir, W.W., 1950, Subsidence of peat lands of the Sacramento-San Joaquin Delta, California: Hilgardia, v. 20, p. 37–55.

FLORIDA EVERGLADES

Allison, R.V., 1956, The influence of drainage and cultivation on subsidence of organic soils under conditions of Everglades reclamation: Soil and Crop Science Society of Florida Proceedings, v. 16, p. 21–31.

Bodle, M.J., Ferriter, A.P., and Thayer, D.D., 1994, The biology, distribution, and ecological consequences of Melaleuca quinquenervia in the Everglades, in Davis, S.M., and Ogden, J.C., The Everglades—The ecosystem and its restoration: Delray Beach, Fla., St. Lucie Press, p. 341–355.

Craft, C.B., and Richardson, C.J., 1993a, Peat accretion and N, P, and organic C accumulation in nutrient-enriched and unenriched Everglades peatlands: Ecological Applications, v. 3, p. 446–458.

Craft, C.B., and Richardson, C.J., 1993b, Peat accretion and phosphorus accumulation along a eutrophication gradient in the northern Everglades: Biogeochemistry, v. 22, p. 133–156.

Davis, J.R., Jr., 1946, The peat deposits of Florida: Florida Geological Survey Bulletin 30, 247 p.

Davis, S.M., and Ogden, J.C., eds., 1994, The Everglades—The ecosystem and its restoration: Delray Beach, Fla., St. Lucie Press, 826 p.

Deren, C.W., Snyder, G.H., Miller, J.D., and Porter, P.S., 1991, Screening for heritability of flood-tolerance in the Florida (CP) sugarcane breeding population: Euphytica, v. 56, p. 155–160.

Douglas, M.S., 1947, The Everglades— River of grass: St. Simons Island,Fla., Mockingbird Press, 308 p.

Gascho, G.J., and Shih, S.F., 1979, Varietal response of sugarcane to water table depth, part 1. Lysimeter performance and plant response: Soil and Crop Society of Florida Proceedings, v. 38, p. 23–27.

Glaz, Barry, 1995, Research seeking agricultural and ecological benefits in the Everglades: Journal of Soil and Water Conservation, v. 50, p. 609–612.

Johnson, Lamar, 1974, Beyond the fourth generation: Gainesville, Fla., The University Presses of Florida, 230 p.

Jones, L.A., Allison, R.V., and others, 1948, Soils, geology, and water control in the Everglades region: University of Florida Agricultural Experiment Station Bulletin 442, 168 p., 4 maps.

Kang, M.S., Snyder, G.H., and Miller, J.D., 1986, Evaluation of Saccharum and related germplasm for tolerance to high water table on organic soil: Journal of the American Society of Sugar Cane Technologists, v. 6, p. 59–63.

Light, S.S., and Dineen, J.W., 1994, Water control in the Everglades: A historical perspective, in Davis, S.M., and Ogden, J.C., The Everglades—The ecosystem and its restoration: Delray Beach, Fla., St. Lucie Press, p. 47–84.

Lucas, R.E., 1982, Organic soils (Histosols)—Formation, distribution, physical and chemical properties and management for crop production: Michigan State University Farm Science Research Report 435, 77 p.

Matson, G.C., and Sanford, Samuel, 1913, Geology and ground water of Florida: U.S. Geological Survey Water-Supply Paper 319, 445 p.

McIvor, C.C., Ley, J.A., and Bjork, R.D., 1994, Changes in freshwater inflow from the Everglades to Florida Bay including effects on biota and biotic processes: A review, in Davis, S.M., and Ogden, J.C., The Everglades—The ecosystem and its estoration: Delray Beach, Fla., St. Lucie Press, p. 117–146.

Ogden, J.C., 1994, A comparison of wading bird nesting colony dynamics (1931–1946 and 1974–1989) as an indication of ecosystem conditions in the southern Everglades, in Davis, S.M., and Ogden, J.C., The Everglades—The ecosystem and its restoration: Delray Beach, Fla.,St. Lucie Press, p. 533–570.

Porter, G.S., Snyder, G.H., and Deren, C.W., 1991, Flood-tolerant crops for low input sustainable agriculture in the Everglades agricultural area: Journal of Sustainable Agriculture, v. 2, p. 77–101.

Ray, J.D., Miller, J.D., and Sinclair, T.R., 1996, Survey of arenchyma in sugarcane roots (abs.): Fifth Symposium of the International Society of Root Research, July 14–18, 1996, Clemson, S. C., p. 118.

Shih, S.F., Glaz, Barry, and Barnes, R.E., Jr., 1997, Subsidence lines revisited in the Everglades agricultural area, 1997: University of Florida Agricultural Experiment Station Bulletin 902, 38 p.

Shih, S.F., Stewart, E.H., Allen, L.H., Jr., and Hilliard, J.E., 1979, Variability of depth to bedrock in Everglades organic soil: Soil and Crop Society of Florida Proceedings, v. 38, p. 66–71.

Smith, G., 1990, The Everglades agricultural area revisited: Citrus and Vegetable Magazine, v. 53, no. 9, p. 40-42.

Smith, T.S., and Bass, O.L., Jr., 1994, Landscape, white-tailed deer, and the distribution of Florida panthers in the Everglades, in Davis, S.M., and Ogden, J.C., The Everglades—The ecosystem and its restoration: Delray Beach, Fla., St. Lucie Press, p. 693–708.

Snyder, G.H., and Davidson, J.M., 1994, Everglades agriculture: Past, present, and future, in Davis, S.M., and Ogden, J.C., The Everglades—The ecosystem and its restoration: Delray Beach, Fla., St. Lucie Press, p. 85–115.

Stephens, J.C., Allen, L.H., Jr., and Chen, Ellen, 1984, Organic soil subsidence, in Holzer, T.L., ed., Man-induced land subsidence: Geological Society of America Reviews in Engineering Geology, v. 6, p. 107–122.

Stephens, J.C., and Johnson, Lamar, 1951, Subsidence of organic soils in the upper Everglades region of Florida: Soil Science Society of Florida Proceedings, v. 11, p. 191–237.

PART III—Collapsing Cavities

INTRODUCTION

Ege, J.R. 1984, Mechanisms of surface subsidence resulting from solution extraction of salt, *in* Holzer, T.L. ed., Man-induced land subsidence: Geological Society of America Reviews in Engineering Geology, v. 6, p. 203–221.

Martinez, J.D., Johnson, K.S., and Neal, J.T., 1998, Sinkholes in evaporite rocks: American Scientist, v. 86, p. 38–51.

White, W.B., Culver, D.C., Herman, J.S., Kane, T.C., and Mylroie, J.E., 1995, Karst lands: American Scientist, v. 83, p. 450–459.

THE RETSOF SALT MINE COLLAPSE, NEW YORK

Alpha Geoscience, 1996, Geologic and hydrogeologic investigation of the Genesee River Valley, prepared for AKZO Nobel Salt Inc., Clarks Summit, Pa.,: Albany, N. Y., Alpha Geoscience Project no. 95132, 31 p., 10 app., 4 plates.

Dunn Corporation, 1992, Hydrogeologic report for the AKZO ash processing plant: Report to Akzo Nobel Salt, Inc., Clarks Summit, Pa., 35 p.

Moran, R.P., Scovazzo, V.A., and Streib, D.L., 1995, Impact analysis—Retsof Mine, Akzo Nobel Salt, Inc.: Report 2455, prepared for the New York State Department of Conservation, J.T. Boyd Co., Inc., 54 p.

Nittany Geoscience, 1995, Groundwater recharge calculations for the Retsof Mine: May 12, 1995, Letter Report, 6 p.

NYSDEC (New York State Department of Environmental Conservation), 1997, Collapse and flooding of Akzo Nobel's Retsof salt mine, Livingston Co., N. Y.: Feb. 1997 Draft report of the Department Task Force, Feb. 1997, 114 p.

Riley, F.S., 1969, Analysis of borehole extensometer data from central California, International Association Of Scientific Hydrology Publication 89, p. 423–431.

Shannon and Wilson, Inc., 1997, Task 3, final report—Retsof Mine collapse, Technical Assistance Grant Committee, Retsof, New York: Seattle, Wash., Shannon and Wilson Inc., 15 p.

Van Sambeek, L.L., 1994, Predicted ground settlement over the Akzo Nobel Retsof Mine, prepared for Akzo Nobel Salt Inc., Clarks Summit, Pa., Project RSI-0525: Rapid City, S. Dak., RE/SPEC Inc., [27+] p.

Van Sambeek, L.L., 1996, Dissolution-induced mine subsidence at the Retsof Salt Mine: Meeting Paper, Solution Mining Research Institute, October 20–23, 1996, Cleveland, Ohio, p. 289–309.

Young, R.A., 1975, The effects of a Late Wisconsin glacial readvance on the postglacial geology of the Genesee Valley, Livingston County, N. Y. [abs.]: Geological Society of America Abstracts with Programs, v. 7, p. 135–136.

SINKHOLES, WEST-CENTRAL FLORIDA

Atkinson, T., 1977, Diffuse flow and conduit flow in limestone terrain in the Mendip Hills, Somerset (Great Britain): Journal of Hydrology, v. 35, p. 93–110.

Bengtsson, T.O., 1987, The hydrologic effects from intense ground-water pumpage in east-central Hillsborough County, Florida, *in* Beck, B.F. and Wilson, W.L. eds, Karst hydrogeology: engineering and environmental applications: proceedings of a conference sponsored by the Florida Sinkhole Research Institute, February 9–11, 1987, College of Engineering, University of Central Florida, Orlando: Boston, Mass., A.A. Balkema, p. 109–114.

Brooks, H.K., 1981, Guide to the physiographic divisions of Florida: Gainesville, Florida Cooperative Extension Service, Institute of Food and Agricultural Sciences, University of Florida, 11 p., 1 plate.

Brucker, R.W., Hess, J.W., and White, W.B., 1972, Role of vertical shafts in the movement of ground water in carbonate aquifers: Ground Water, v. 10, p. 5–13.

Culshaw, M.G., and Waltham, A.C., 1987, Natural and artificial cavities as ground engineering hazards: Quarterly Journal of Engineering Geology, v. 20. p. 139–150.

Ford, D., and Williams, P., 1989, Karst Geomorphology and Hydrology: Boston, Mass., Unwin Hyman, 601 p.

Lattman, L.H., and Parizek, R.R., 1964, Relationship between fracture traces and the occurrence of ground water in carbonate rocks: Journal of Hydrology, v. 2, p. 73–91.

Lewelling, B. R., Tihansky, A. B., and Kindinger, J. L., 1998, Assessment of the hydraulic connection between ground water and the Peace River, west-central Florida: U. S. Geological Survey Water-Resources Investigation Report 97-4211, 96 p.

Littlefield, J.R., Culbreth, M.A., Upchurch, S.B., and Stewart, M.T., 1984, Relationship of modern sinkhole development to large-scale photolinear features: Multidisciplinary Conference on Sinkholes, 1st, Orlando, Fla., October 15–17, [Proceedings, Beck, B.F., ed., Sinkholes—Their geology, engineering and environmental impact: Boston, Mass., A.A. Balkema], p. 189–195.

Metcalfe, S.J., and Hall, L.E., 1984, Sinkhole collapse due to groundwater pumpage for freeze protection irrigation near Dover, Florida, January, 1977: Multidisciplinary Conference on Sinkholes, 1st, Orlando, Florida, October 15–17, [Proceedings, Beck, B.F., ed., Sinkholes—Their geology, engineering and environmental impact: Boston, Mass., A.A. Balkema], p. 29–33.

Mularoni, R. A., 1993, Potentiometric surface of the Upper Floridan Aquifer, west-central Florida, September 1992: U. S. Geological Survey Open-File Report 93-49, 1 plate.

Newton, J.G., 1986, Development of sinkholes resulting from man's activities in the eastern United States: U.S. Geological Survey Circular 968, 54 p.

Quinlan, J.F., Davies, G.J., and Worthington, S.R.,1993, Review of groundwater quality monitoring network design: Journal of Hydraulic Engineering, v. 119, p. 1,436–1,441. [Discussion, with reply, p. 1,141–1,142]

Ryder, P.D., 1985, Hydrology of the Floridan aquifer system in west-central Florida: U.S. Geological Survey Professional Paper 1403-F, 63 p., 1 plate.

Sinclair, W.C., 1982, Sinkhole development resulting from ground-water development in the Tampa area, Florida: U.S. Geological Survey Water-Resources Investigations Report 81-50, 19 p.

Sinclair, W.C., and Stewart, J.W., 1985, Sinkhole type, development, and distribution in Florida: U.S. Geological Survey Map Series 110, 1 plate.

Southeastern Geological Society, 1986, Hydrogeological Units of Florida: Florida Geological Survey Special Publication 28, 9 p.

Sowers, G.F., 1984, Correction and protection in limestone terrane: Multidisciplinary Conference on Sinkholes, 1st, Orlando, Florida, October 15–17, [Proceedings, Beck, B.F., ed., Sinkholes—Their geology, engineering and environmental impact: Boston, Mass., A.A. Balkema], p. 373–378.

Stewart, M., and Parker, J., 1992, Localization and seasonal variation of recharge in a covered karst aquifer system, Florida, USA: International Contributions to Hydrogeology, v. 13, Springer-Verlag, p. 443–460.

Tihansky, A.B., and Trommer, J. T., 1994, Rapid ground-water movement and transport of nitrate within a karst aquifer system along the coast of west-central Florida [abs.]: Transactions American Geophysical Union, v. 75, April 19, 1994—Supplement, p. 156.

Trommer, J.T., 1992, Effects of effluent spray irrigation and sludge disposal on ground water in a karst region, northwest Pinellas County, Florida: U.S. Geological Survey Water-Resources Investigations Report 91-4181, 32 p.

Watts, W.A., 1980, The Late Quaternary vegetation history of the southeastern United States: Annual Review of Ecology and Systematics, v. 11, p. 387–409.

Watts, W.A., and Stuiver, M., 1980, Late Wisconsin climate of northern Florida and the origin of species-rich deciduous forest: Science, v. 210, p. 325–327.

Watts, W.A., and Hansen, B.C.S., 1988, Environments of Florida in the Late Wisconsinan and Holocene, in Purdy, B.A., ed., Wet site archeology: Caldwell, N.J., Telford West, p. 307–323.

White, W.A.,1970, The geomorphology of the Florida peninsula: Florida Bureau of Geology Geological Bulletin 51, 164 p.

Wilson, W.L., and Shock, E.J., 1996, New sinkhole data spreadsheet manual (v1.1): Winter Springs, Fla., Subsurface Evaluations, Inc., 31 p., 3 app., 1 disk.

The Role of Science

Amelung, F., Galloway, D.L., Bell, J.W., Zebker, H.A., and Laczniak, R.J., 1999, Sensing the ups and downs of Las Vegas—InSAR reveals structural control of land subsidence and aquifer-system deformation: Geology, v. 27, p. 483–486.

Anderson S.R., 1988, Potential for aquifer compaction, land subsidence, and earth fissures in Tucson Basin, Pima County, Arizona: U.S. Geological Survey Hydrologic Investigations Atlas HA-713, 3 sheets, scale 1:250,000.

Anderson, S.R., 1989, Potential for aquifer compaction, land subsidence, and earth fissures in Avra Valley, Pima and Pinal Counties, Arizona: U.S. Geological Survey Hydrologic Investigations Atlas HA-718, 3 sheets, scale 1:250,000.

Bear, J., 1979, Hydraulics of groundwater: New York, McGraw-Hill, 569 p.

Biot, M.A., 1941, General theory of three-dimensional consolidation: Journal of Applied Physics, v. 12, p. 155–164.

Blomquist, W., 1992, Dividing the waters—Governing groundwater in southern California: San Francisco, Calif., ICS Press, 413 p.

Carpenter, M.C., 1993, Earth-fissure movements associated with fluctuations in ground-water levels near the Picacho Mountains, south-central Arizona, 1980–84: U.S. Geological Survey Professional Paper 497-H, 49 p.

Fielding, E.J., Blom, R.G., and Goldstein, R.M., 1998, Rapid subsidence over oil fields measured by SAR interferometry: Geophysical Research Letters, v. 27, p. 3,215–3,218.

Gabriel, A.K., Goldstein, R.M., and Zebker, H.A., 1989, Mapping small elevation changes over large areas—Differential radar interferometry: Journal of Geophysical Research, v. 94, p. 9,183–9,191.

Galloway, D.L., Hudnut, K.W., Ingebritsen, S.E., Phillips, S.P., Peltzer, G., Rogez, F., and Rosen, P.A., 1998, Detection of aquifer system compaction and land subsidence using interferometric synthetic aperture radar, Antelope Valley, Mojave Desert, California: Water Resources Research, v. 34, p. 2,573–2,585.

Hanson, R.T., 1989, Aquifer-system compaction, Tucson Basin and Avra Valley, Arizona: U.S. Geological Survey Open-File Report 88-4172, 69 p.

Hanson, R.T., Anderson, S.R., and Pool, D.R., 1990, Simulation of ground-water flow and potential land subsidence, Avra Valley, Arizona: U.S. Geological Survey Water-Resources Investigations Report 90-4178, 41 p.

Hanson, R.T., and Benedict, J.F., 1994, Simulation of ground-water flow and potential land subsidence, upper Santa Cruz Basin, Arizona: U.S. Geological Survey Water-Resources Investigations Report 93-4196, 47 p.

Helm, D.C., 1975, One-dimensional simulation of aquifer system compaction near Pixley, Calif., part 1. Constant parameters: Water Resources Research, v. 11, p. 465–478.

Helm, D.C., 1978, Field verification of a one-dimensional mathematical model for transient compaction and expansion of a confined aquifer system: American Society of Civil Engineers Hydraulics Division Specialty Conference, 26th, University of Maryland, College Park, Md., August 9–11, 1978, p. 189–196.

Heywood, C.E., 1995, Investigation of aquifer-system compaction in the Hueco basin, El Paso, Texas, USA: International Symposium on Land Subsidence, 5th, Delft, Netherlands, October 1995, International Association of Hydrological Sciences Publication 234, p. 35–45.

Holzer, T.L., 1981, Preconsolidation stress of aquifer systems in areas of induced land subsidence: Water Resources Research, v. 17, p. 693–704.

Ikehara, M.E., Galloway, D.L., Fielding, E., Bürgmann, R., Lewis, A.S., and Ahmadi, B., 1998, InSAR imagery reveals seasonal and longer-term land-surface elevation changes influenced by ground-water levels and fault alignment in Santa Clara Valley, California [abs.]: EOS (supplement) Transactions, American Geophysical Union, no. 45, November 10, 1998, p. F37.

Ikehara, M.E., and Phillips, S.P., 1994, Determination of land subsidence related to ground-water level declines using global positioning system and leveling surveys in Antelope Valley, Los Angeles and Kern Counties, California, 1992: U.S. Geological Survey Water-Resources Investigations Report 94-4184, 101 p.

Massonnet, D., Briole, P., and Arnaud, A., 1995, Deflation of Mount Etna monitored by spaceborne radar interferometry: Nature, v. 375, p. 567–570.

Massonnet, D., and Feigl, K.L., 1998, Radar interferometry and its application to changes in the earth's surface: Reviews of Geophysics, v. 36, p. 441–500.

Massonnet, D., Holzer, T., and Vadon, H., 1997, Land subsidence caused by the East Mesa geothermal field, California, observed using SAR interferometry: Geophysical Research Letters, v. 24, p. 901–904.

Massonnet, D., Rossi, M., Carmona, C., Adragna, F., Peltzer, G., Feigl, K., and Rabaute, T., 1993, The displacement field of the Landers earthquake mapped by radar interferometry: Nature, v. 364, p. 138–142.

National Research Council, 1991, Mitigating losses from land subsidence in the United States: Washington, D. C., National Academy Press, 58 p.

Riley, F.S., 1969, Analysis of borehole extensometer data from central California, International Association of Scientific Hydrology Publication 89, p. 423–431.

Riley, F.S., 1986, Developments in borehole extensometry: International Symposium on Land Subsidence, 3rd, Venice, 19–25 March 1984, [Proceedings, Johnson, I.A., Carborgnin, Laura, and Ubertini, L., eds.], International Association of Scientific Hydrology Publication 151, p. 169–186.

Rosen, P.A., Hensley, S., Zebker, H.A., Webb, F.H., and Fielding, E., 1996, Surface deformation and coherence measurements of Kilauea volcano, Hawaii, from SIR-C radar interferometry: Journal of Geophysical Research, v. 101, p. 23,109–23,125.

Tihansky, A.B., Arthur, J.D., and DeWitt, D.J., 1996, Sublake geologic structure from high-resolution seismic-reflection data from four sinkhole lakes in the Lake Wales Ridge, Central Florida: U.S. Geological Survey Open-File Report 96-224, 72 p.

Terzaghi, K., 1925, Principles of soil mechanics, IV—Settlement and consolidation of clay: Engineering News-Record, v. 95, p. 874–878.

Vadon, H., and Sigmundsson, F., 1997, 1992–1995 crustal deformation at Mid-Atlantic ridge, SW Iceland, mapped by radar interferometry: Science, v. 275, p. 193–197.

Wicks, C., Jr., Thatcher, W., and Dzurisin, D., 1998, Migration of fluids beneath Yellowstone Caldera inferred from satellite radar interferometry: Science, v. 282, p. 458–462.

Zebker, H.A., Rosen, P.A., Goldstein, M., Gabriel, A., and Werner, C.L., 1994, On the derivation of coseismic displacement fields, using differential radar interferometry—The Landers earthquake: Journal of Geophysical Research, v. 99, p. 19,617–19,634.

Zilkoski, D.B., D'Onofrio, J.D., and Frakes, S.J., 1997, Guidelines for establishing GPS-derived ellipsoid heights (Standards: 2 cm and 5 cm), ver. 4–3: National Oceanic and Atmospheric Administration Technical Memorandum NOS NGS-58, [20+] p.

United States Geological Survey

SELECTED PUBLICATIONS AND AVAILABILITY

PUBLICATIONS

Books and other publications

Professional Papers report scientific data and interpretations of lasting scientific interest that cover all facets of USGS investigations and research.

Bulletins contain significant data and interpretations that are of lasting scientific interest but are generally more limited in scope or geographic coverage than Professional Papers.

Water-Supply Papers are comprehensive reports that present significant interpretive results of hydrologic investigations of wide interest to professional geologists, hydrologists, and engineers. The series covers investigations in all phases of hydrology, including hydrogeology, availability of water, quality of water, and use of water.

Circulars are reports of programmatic of scientific information of an ephemeral nature; many present important scientific information of wide popular interest. Circulars are distributed at no cost to the public.

Fact Sheets communicate a wide variety of timely information on USGS programs, projects, and research. They commonly address issues of public interest. Fact sheets generally are two or four pages long and are distributed at no cost to the public.

Reports in the **Digital Data Series (DDS)** distribute large amounts of data through digital media, including compact disc-read-only memory (CD-ROM). They are high-quality, interpretive publications designed as self-contained packages for viewing and interpreting data and typically contain data sets, software to view the data, and explanatory text.

Water-Resources Investigations Reports are papers of an interpretive nature made available to the public outside the formal USGS publications series. Copies are produced on request (unlike formal USGS publications) and are also available for public inspection at depositories indicated in USGS catalogs.

Open-File Reports can consist of basic data, preliminary reports, and a wide range of scientific documents of USGS investigations. Open-File Reports are designed for fast release and are available for public consultation at depositories.

Maps

Geologic Quadrangle Maps (GQs) are multicolor geologic maps on topographic bases in 7.5- or 15-minute quadrangle formats (scales mainly 1:24,000 or 1:62,500) showing bedrock, surficial, or engineering geology. Maps generally include brief texts; some maps include structure and columnar section only.

Geophysical Investigations Maps (GPs) are on topographic or planimetric bases at various scales. They show results of geophysical investigations using gravity, magnetic, seismic, or radioactivity surveys, which provide data on subsurface structures that are of economic or geologic significance.

Miscellaneous Investigations Series Maps or **Geologic Investigations Series (Is** are on planimetric or topographic bases at various scales; they present a wide variety of format and subject matter. The series also includes 7.5-minute quadrangle photogeologic maps on planimetric bases and planetary maps.

Information Periodicals

Metal Industry Indicators (MIIs) is a free monthly newsletter that analyzes and forecasts the economic health of five metal industries with composite leading and coincident indexes: primary metals, steel, copper, primary and secondary aluminum, and aluminum mill products.

Mineral Industry Surveys (MISs) are free periodic statistical and economic reports designed to provide timely statistical data on production, distribution, stocks, and consumption of significant mineral commodities. The surveys are issued monthly, quarterly, annually, or at other regular intervals, depending on the need for current data. The MISs are published by commodity as well as by State. A series of international MISs is also available.

Published on an annual basis, **Mineral Commodity Summaries** is the earliest Government publication to furnish estimates covering nonfuel mineral industry data. Data sheets contain information on the domestic industry structure, government programs, tariffs, and 5-year salient statistics for more than 90 individual mineral and materials.

The Minerals Yearbook discusses the performance of the worldwide minerals and materials industry during a calendar year, and it provides background information to assist in interpreting that performance. The Minerals Yearbook consists of three volumes. Volume I, Metals and Minerals, contains chapters about virtually all metallic and industrial mineral commodities important to the U.S. economy. Volume II, Area Reports: Domestic, contains a chapter on the minerals industry of each of the 50 states and Puerto Rico and the Administered Islands. Volume III, Area Reports: International, is published as four separate reports. These reports collectively contain the latest available mineral data on more than 190 foreign countries and discuss the importance of minerals to the economies of these nations and the United States.

Permanent Catalogs

"Publications of the U.S. Geological Survey, 1879–1961" and **"Publications of the U.S. Geological Survey, 1962–1970"** are available in paperback book form and as a set of microfiche.

"Publications of the U.S. Geological Survey, 1971–1981" is available in paperback book form (two volumes, publications listing and index) and as a set of microfiche.

Annual supplements for 1982, 1983, 1984, 1985, 1986, and subsequent years are available in paperback book form.

AVAILABILITY OF PUBLICATIONS

Order U.S. Geological Survey (USGS) publications by calling the toll-free telephone number 1-888-ASK-USGS or contacting the offices listed below. Detailed ordering instructions, along with prices of the last offerings, are given in the current-year issues of the catalog "New Publications of the U.S. Geological Survey."

Books, Maps, and Other Publications

By Mail

Books, maps and other publications are available my mail from:

USGS Information Series
Box 25286, Federal Center
Denver, CO 80225

Publications include Professional Papers, Bulletins, Water-Supply Papers, Techniques of Water-Resources Investigations, Circulars, Fact Sheets, publications of general interest, single copies of permanent USGS catalogs, and topographic and thematic maps.

Over the Counter

Books, maps, and other publications of the U.S. Geological Survey are available over the counter at the following USGS Earth Science Information Centers (ESICs), all of which are authorized agents of the Superintendent of Documents:

- Anchorage, Alaska—Rm. 101, 4230 University Dr.
- Denver, Colorado—Bldg. 810, Federal Center
- Menlo Park, California—Rm. 3128, Bldg. 3, 345 Middlefield Rd.
- Reston, Virginia—1C402, USGS National Center, 12201 Sunrise Valley Dr.
- Salt Lake City, Utah—2222 West, 2300 South
- Spokane, Washington—Rm. 135, U.S. Post Office Building, 904 West Riverside Ave.
- Washington, D.C.—Rm. 2650, Main Interior Bldg., 18th and C Sts., NW

Maps only may be purchased over the counter at the following USGS office:

- Rolla, Missouri—1400 Independence Rd.

Electronically

Some USGS publications, including the catalog "New Publications of the U.S. Geological Survey" are also available electronically on the USGS's World Wide Web home page at **http://www.usgs.gov**

Preliminary Determination of Epicenters

Subscriptions to the periodical "Preliminary Determination of Epicenters" can be obtained only from the Superintendent of Documents. Check or money order must be payable to the Superintendent of Documents. Order by mail from:

Superintendent of Documents
Government Printing Office
Washington, DC 20402

Information Periodicals

Many Information Periodicals products are available through the systems or formats listed below:

Printed Products

Printed copies of the Minerals Yearbook and the Mineral Commodity Summaries can be ordered from the Superintendent of Documents, Government Printing Office (address above). Printed copies of Metal Industry Indicators and Mineral Industry Surveys can be ordered from the Center for Disease Control and Prevention, National Institute for Occupational Safety and Health, Pittsburgh Research Center, P.O. Box 18070, Pittsburgh, PA 15236-0070

Mines FaxBack: Return fax service

1. Use the touch-tone handset attached to your fax machine's telephone jack. (ISDN [digital] telephones cannot be used with fax machines.)

2. Dial (703) 648-4999

3. Listen to the menu options and punch in the number of your selection, using the touch-tone telephone.

4. After completing your selection, press the start button on your fax machine.

CD-ROM

A disc containing chapters of the Minerals Yearbook (1993–95), the Mineral Commodity Summaries (1995–97), a statistically compendium (1970–90), and other publications is updated three times a year and sold by the Superintendent of Documents, Government Printing Office (address above).

World Wide Web

Minerals information is available electronically at **http://minerals.er.usgs.gov/minerals/**

Subscription to the catalog "New Publications of the U.S. Geological Survey"

Those wishing to be placed on a free subscription list for the catalog "New Publications of the U.S. Geological Survey" should write to:

U.S. Geological Survey
903 National Center
Reston, VA 20192